PRAGUE

Edited by Joachim Chwaszcza
Photography by Bodo Bondzio and Joachim Chwaszcza
Designed and Directed by Hans Johannes Hoefer

APA PUBLICATIONS

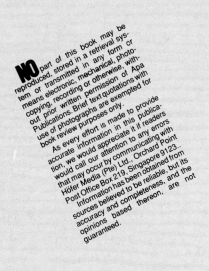
PRAGUE

First Edition
© 1989 APA PUBLICATIONS (HK) LTD
All Rights Reserved
Printed in Singapore by Höfer Press Pte. Ltd

ABOUT THIS BOOK

Cityguide: Prague continues the series by APA Publications on "the world's most beautiful cities". This new, but already successful, series grew out of the prize-winning *Insight Guides,* developed in style and concept by **Hans Hoefer**, the founder of APA Publications whose first laurel was the famous *Insight Guide: Bali*, published in 1970.

Cityguide: Prague has the same striking combination of photography and text, useful travel information and up-to-date maps.

Prague, the former European metropolis, is today at the interface between East and West, a place full of fascinating relics of a splendid and turbulent past, with its myriad squares, streets and alleys. Prague is also the birthplace of such respected works as Rabbi Löw's *Golem*, Hasek's *Good Soldier Schwejk*, and Kafka's *Joseph K.*

A Close Look

Cityguide: Prague focusses on the city's architecture and art history, and delves into the daily life of the city and, of course, its interesting aspects. Special features on the little guided tour of avant-garde architecture between 1900 and 1960, and of Prague literary life, reveal that though the city may not offer much exotic excitement to the traveler, it has lots of charm.

Joachim Chwaszcza, a freelance photographer and writer from Munich, planned, edited and produced the book. The book has a special meaning for him; it takes him back to his own history as one of his parents came from Czechoslovakia. Chwaszcza also contributed several articles and pictures to the book, and, with the help of his many Prague friends, compiled the Travel Tips.

The Authors

Chwaszcza's sister, Christine Chwaszcza is studying political science in Munich and is close to getting her degree. She compiled, with almost scientific care, the first part of the historical introduction, and the early humanist circle around Charles IV.

Eva Meschede also studied political science in Munich, attended a school for journalists and works freelance for newspapers and the radio.

Ota Filip was born in Czechoslovakia and was educated at the Academy for Journalism. Filip, who is a member of PEN, has been living in Munich since 1975, working freelance as a writer and author. He has written 10 novels, the latest of which is *Cafe Slavia*. His works have been translated into seven languages.

Vilem Wagner was born in Prague and lived there till he was 21. He studied the violin and music in Prague, Munich and Hamburg and has worked in films and television. Wagner lives in Hamburg, but he remains Czech in his heart and soul.

Johanna von Herzogenberg was born in Sichrow, North Bohemia. She studied German and Art History in Prague and Tübingen and graduated from Charles University in Prague. She has written for a number of art history publications. She works for newspapers and radio, and arranges exhibitions in the United States, Israel, Austria, and the Federal Republic of Germany.

Jana Kubalova, director in the Glass and Ceramics section of the Craft Museum in

J. Chwaszcz   Ch. Chwaszcza   Meschede   Filip   Wagner

Prague, contributed a short history of Bohemian glass.

Dr. Frantisek Kafka is perhaps the most important person in Prague. He was a civil servant in the first post-war government; from 1947 onward, he was a member of the PEN Club. He was leader of the Jewish community in Prague from 1974 to 1978. Kafka has had more than 20 books published, and still lives an active life as a writer.

Franz Peter Künzel was one of the first to brave the Cold War and editorial caution to popularize Czech literature in the West. He translated Hrbal and many other well-known authors and received the Translators' Prize from the Czech Writers Union.

Marc Rehle works as an architect. He lectures in Design and Preservation of Historic Monuments at the Technische Universität in Munich. His interest in the countries of Eastern Europe has continued to grow since his student years, and he has explored Czechoslovakia and other East European countries during frequent visits in his on-going search for important but forgotten 20th-century buildings.

The Photographers

The photographs in *Cityguide: Prague* bear the hallmark of **Bodo Bondzio,** who is married to a native of Prague and has taken photos of the city for nearly 10 years. In that time, there can be few places in the city he hasn't covered. Graphic designer Bondzio provided most of the material for the guide. Supplementary photos were shot by J. Chwaszcza and **Jens Schumann. Pavel Scheufler** provided several photos from his rich archives of historic pictures—a collection accumulated over 20 years.

Extraordinary Efforts

Sincere thanks go to **L.** and **P. Vilas, J. Kafka** and **L. Adamek** for their invaluable support.

Special thanks to free-lance writer and translator **Susan James** who translated the German text of *Prague*. James was born in West Germany and spent much of her childhood in West Germany and England. She has worked as a teacher, researcher, proofreader, journalist, magazine editor and translator. She also runs a commercial photography studio with her brother.

—APA PUBLICATIONS

Herzogenberg

Kafka

Künzel

Rehle

Bondzio

CONTENTS

TRAVEL TIPS

THE FORGOTTEN JEWEL IN THE BOHEMIAN CROWN

Prague is certainly not at the top of the list of the great cities of Europe. In the papers, the amount of news about Prague is small and it gets less all the time. If we happen to think of something in connection with Prague, we usually think in terms of catch-phrases such as "Golden Prague", "Prague of the Hundred Spires", and possibly sometimes the "Prague Spring".

"History hasn't been made here for quite a long time", claims one novel about Prague. Is Prague a mere provincial town then, far away from the places where today's decisions are taken and changes are made? Or is it, like so many things that aren't in the limelight, an insider's tip? Every visitor should be allowed to find that out for him or herself.

Is Prague then nothing but a giant open-air museum, a city almost undamaged by the last war? Nothing but the medieval castle, Baroque churches and the palaces of the nobility? Or is it a paradise for Westeners with currency exchanged on the black market, for coachloads of pensioners or noisy schoolchildren, perhaps even nothing but the goal of shopping expeditions from the provinces or the less well-off neighbors in East Germany?

It's probably a bit of everything, and a little bit more as well. Prague is still a Bohemian city—in the artistic sense. The former restful lifestyle, the coffee house idyll, has tried to move with the times. This is a city that loves life and perhaps isn't so far away from old imperial times as it would have us believe. Is it a melancholy, perhaps even slightly morbid breeze that blows through the narrow streets and coffee houses?

This is a city that lives off its stories—new ones and old ones. Anyone who knows or gets to know the people of Prague will also learn how important these little anecdotes are. Prague is a city of storytellers, for everyone has a story to tell. Stories, anecdotes, jokes—people use these to express the things that oppress them.

Prague is not a modish city, and certainly not a modern one. To be sure, you can see modern buildings, there are even some very recent developments to be proud of. But there is no pulsating life in the streets, there is scarcely anything resembling an "in-crowd" (and if there is, it's hidden away and hardly accessible to outsiders). Many of the best people have gone abroad, for a long time or permanently. This is a great loss which has much impoverished Prague. The number of those who have left or have been driven out is too great.

Don't worry about propaganda from either side. That is probably one of the best pieces of advice anyone could give a traveler about to set off for Prague. Also: "Prague is magic, something that ties you down and holds you and always draws you back. You can never forget", wrote the author and journalist Egon Erwin Kisch.

Preceding pages: Prague c. 1860; young woman of Prague; Charles Bridge; Dome of St. Nicholas in the Malá Strana. Left, the Old Town Hall.

LIBUSSA AND THE BOHEMIAN KINGS

The founding of Prague is surrounded by legend. The chronicler Cosmas tells how Libussa, the wife of Přemysl, persuaded her husband to search out an unimpressive village on the banks of the Vltava and found a city there. She prophesied great things for the city: "The time will come when two golden olive trees will grow in this city. Their tops will reach the seventh heaven and they will shine throughout the world through signs and wonders."

According to the legend, Přemysl and his followers went to the place Libussa described and there founded Prague.

Rise of Přemyslid Dynasty

Archaeological finds show that the area around Prague has been inhabited since neolithic times. However, the evolution of Prague as the political and cultural center of Bohemia is very closely bound up with the rise and the power of the Přemyslid dynasty. The battle for supremacy in the Bohemian and Moldavian regions between the Slavnik and the Přemyslid groups was won by the Přemyslids towards the end of the 9th century. Their rulers moved their residence from Levy Hradec to the strategically useful rocky outcrop on the left bank of the Vltava and built a castle there.

The etymological origin of the name "Praha", which at first only referred to the Přemyslid castle, is not known for certain, but appears to be connected with the aridity of the region. Settlements developed in the area at the front of the castle. At first these were inhabited only by people directly supplying the castle, but later craftsmen also settled here. In the early 10th century the Vyšehrad (originally known as the "Chrasten") was built, on the left bank of the Vltava some 2 1/2 miles (4 km) south of the old castle. The Hradčany was built at about the same time. Under Boleslav I (929-967 or

972) Bohemia was finally forced to join the Holy Roman Empire, and after the expulsion of the Magyars, Bohemia cultivated its connections to the West, extending them as far as Rome. Prague became an important trading center. In 965, the merchant Ibrahim Ibn Yakub wrote that "the city of Prague is built of stone and lime and is the greatest trading city". This might be a slight exaggeration, the clay walls of the castle were probably not replaced with stone until the 12th century.

Founding of Bishopric of Prague

A decisive point in the development of Prague was the success of Boleslav II (967/72-999) in obtaining the consent, long withheld, of the Holy Roman Emperor for the founding of a bishopric in Prague in 973. The first bishop was the monk Adalbert, trained in the cathedral school in Magdeburg (an important center of the Slav Mission) and appointed in 982. Adalbert fled from Prague on several occasions and died in Poland. His relics were seized by Břetislav I (1040-1055) during a campaign in Poland and brought back to Prague. The plan to use the relics to raise the status of the Prague cathedral came to nothing, but Prague increased in international importance during Břetislav's reign. Břetislav's successor Spytihnev II (1055-1061) is credited with the expulsion of German merchants from Prague and with the building of a Romanesque St. Vitus' Basilica on the St. Vitus' Rotunda in the castle of Prague. His successor Duke Vratislav I (1061-1092) proved himself a true ally of the Emperor Henry IV, who, at the Synod of Mainz in 1085, appointed him King of Bohemia. The coronation was held in Prague in 1086 by Engelbert, Archbishop of Trier.

Marketplace of Prague

The marketplace of Prague is first mentioned in documents dating from the late 11th century. It was situated in the area of today's Staroměstské náměsti (Old Town

Square). In the early 12th century the settlement in front of the castle shifted to the right bank of the Vltava. An incredible construction boom now began, and the Romanesque style spread throughout Prague. Sobeslav I (1125-1140) completed the alterations to the Vyšehrad, begun by Vratsilav II. The Romanesque churches and the castle citadel were completed and the clay walls replaced with stone fortifications. Sobeslav did all he could to encourage the expansion of the economy and of trade. Under his patronage, the Vyšehrad Cathedral Chapter produced the Vyšehrad Codex, a richly decorated manuscript. Vladislav II (1140-1172—known as Vladislav I after his coronation as King of Bohemia) moved the ruler's residence back to the old castle. The new king had an aristocratic palace built, and the cathedral of St. Vitus was extended. In 1140, the Premonstratensian monastery on the Strahov was built, and in 1170 the St. John's monastery was founded on the elevation of what was later to be the Malá Strana bridgehead of the Judith Bridge.

Settlement outside the castles also increased considerably, as the many church foundations dating from this time show—St. John's, Holy Cross, St. Laurence, St. Andrew, St. Leonard, St. Mary, SS Philip and James in the Bethlehem Square, St. Clement and St. Aegidius, almost all of them in the Old Town. King Přemysl Otokar III (1197-1230) founded another "new town" in today's Mala Strana (Lesser Quarter—literally "small side") just under the castle. He also completed the building of the stone Judith Bridge (about the same level as today's Charles Bridge), begun in 1170, and created a permanent link between the castle and the Old Town. Under Otokar's rule, Bohemia rose to become an important Central European power. Prague, residence of the Přemyslid dynasty, was promoted to one of the international meeting places of Central Europe. Long-distance trade had largely been re-routed along the Danube and through Vienna, but local needs and the luxury trade made up for the shortfall.

By 1231, at the latest, King Wenceslas I (1230-1253) had begun the fortification of the Old Town and the newly founded settlement of Havelské Mesto. The building of a city wall was a sign of extraordinary privilege, as a wall meant more than protection against attacks from outside. A slogan of the times was "city air makes for freedom". The urban population was not formed of serfs and was not tied to the land. If you lived in a town for a year, you became one of the citizens. The Old Town was still under the jurisdiction of a judge appointed by the king—his position was equivalent to that of a governor—but Havelské Mesto was granted rights similar to those in force in the free cities of the Holy Roman Empire.

The city was administered by a council which met in the house of the judge. Not until c. 1338 did John of Luxembourg grant the Old Town the right to its own town hall. The city council bought the house of Wölflin vom Stein, and later added the neighboring house. The city judge and the councillors were nominated from among the patrician families (nobility and land-owners) and elected by the members of those same families. It is assumed that the first guilds and craft associations arose in Prague towards the end of the 13th century. The Jewish quarter was ruled directly by the king, and some church property kept its special status.

At this time, the city wall was laid out mainly according to strategic considerations. By 1287 at the latest, when the laws of the Old Town were officially current in Havelské Mesto as well, we can assume a consolidation of the city. The older settlement (Hradčany) on the right bank received a city charter and walls in 1320, the Malá Strana (known as "new town" in those days) was granted privileges and rights equivalent to the Magdeburg Charter in 1257.

At the beginning of the 14th century, then, Prague was composed of three different towns, each legally and administratively independent and also very different in their social and demographic structure. The total area of the city in 1300 is estimated at 297 acres (120 ha), the Old Town alone at 198 acres (80 ha).

THE IMPERIAL CAPITAL

End of Přemyslid Dynasty

At the beginning of the 14th century the rule of the Přemyslids in Bohemia ended. Wenceslas III died while still a child. For a while Henry of Carinthia, Wenceslas II's brother-in-law, ruled, then it was the turn of Rudolf of Habsburg, then Henry again.

The times were marked by unrest and political power struggles, and the patrician families of Prague took part in them, with varied success. The city was besieged, laid to waste and plundered several times. In 1310, the Bohemian Estates chose John of Luxembourg (1296-1346) as their new king. On Aug. 31, 1310 he married Elizabeth, the younger sister of Wenceslas II, and on the same day he received Bohemia from his father, the Emperor Henry VII. John was intensely—and expensively—involved in imperial politics and spent most of his time out of the country. The castle of Prague fell gradually into decay, but the city was able to buy numerous privileges from John, among them the town hall mentioned above.

Emperor Charles IV

However, John's son, Charles IV, was particularly fond of Prague. Charles was born in 1316 and baptized Wenceslas, but at his confirmation he took Charlemagne as his personal patron and changed his name. In 1323, John brought young Wenceslas to Paris, where Petrus Rogerii—later Pope Clement VI—was entrusted with his education. In 1333, Charles, after a short period spent in Italy, came back to Prague as his father's governor, and had the dilapidated castle rebuilt according to French models. However, Charles did not immediately take up the regency (with the recognition of the Bohemian Estates) of the Bohemian crown lands untill 1340. He was now acting for his

Preceding pages: extract from the "Golden Bull" of Charles IV. Left, Charles IV, from a votive tablet by Johann Ocko.

father, who had gone blind. In 1347 he was crowned King of Bohemia and in 1349—after the death of his rival Louise the Bavarian—he was elected Holy Roman Emperor for the second time.

He chose Prague to be his imperial residence and spared no effort to expand it to be not only the political, but also the cultural hub of Central Europe.

Peter Parler—Imperial Architect

In 1344, Charles had already managed to use his good relationship with Clement IV to get Prague promoted from a bishopric to an archbishopric. In the same year the building of St. Vitus' Cathedral began, over the remains of the former basilica. The cathedral was conceived as a three-aisled nave cathedral church, which has French influence. Charles obtained the services of a very famous architect, Matthias of Arras.

Following the latter's death in 1352, the masons' and sculptors' workshop of Peter Parler (1332/33-1399) took over the building. Under his direction the famous triforium arcade was created, along with the choir, the South Tower, and particularly the vaulting, for which Parler was famous. The Wenceslas Chapel and the Golden Gate, both evidence of a synthesis of architecture and sculpture never before achieved, can be attributed wholly to Parler. After finishing work on St. Vitus' Cathedral, Parler was contracted to complete the church of St. Mary in the Teyn. Also designed and executed by Parler's workshop is the work around the windows of the Martinic chapel, the Bridge Tower of the Old Town and the church of the Charles Church, which was modeled on the palace chapel of Charlemagne in Aachen. In 1357, under Parler's direction, work began on the Charles Bridge.

Peter Parler was probably the most notable architect of his time. His style, which was widely imitated, clearly foreshadows later developments made in architecture during Renaissance times.

Prague gained considerably in cultural importance when the university—the first in Central Europe—was built. Charles IV granted the official founding Charter on Apr. 7, 1348. The Charles University was intended to draw together scholars from all the regions of the empire and had a constitution similar to that of the University of Paris, where Charles had studied. It was divided into four "nations", Bohemian, Bavarian, Saxon and Polish. However, these did not represent actual national groupings, but symbolized the four points of the compass by giving them the name of the nearest national cultural centers of Europe. Among the early Humanist circle around Charles were people like Cola di Rienzi, Ernest of Pardubitz (later Archbishop of Prague) and the famous Johann of Neumark. He was Charles' chancellor and had considerable influence on the spread of the New High German language. He was famous as a translator of Latin prayers and Bible texts.

A Masterpiece of Civic Planning

However, Charles made his most noticeable mark on the development of Prague by

neighbor in that direction. The "nations" were important insofar as each had a vote in the University's decision-making process, and the posts of Rector and Chancellor were filled by each "nation" in turn.

To begin with, lectures were held in churches and in the "Lazarus House" in the Jewish Quarter. The move to the Carolinum did not take place until 1356. Charles, who was a writer himself and was one of the few educated medieval rulers, was able to gather leading thinkers and scholars of his time around him. Without doubt, Prague in the 14th century was one of the most important

founding the New Town *(Nové mesto)* and thereby almost doubling its area. The New Town spread in a circle from the southeast of the Old Town to the river below the Vyšehrad and to the Porici. This area incorporated some smaller settlements and several monasteries, among them the Carmelite convent with the church of St. Mary of the Snows and the Emmaus monastery, which were founded in 1347 and already planned to fit in with the designs for the future New Town. Also in 1347, the foundation stone of the Charles Church was laid.

This same year also saw the first announcements of the founding of a new town, the site of which had to be "thoroughly considered and well advised". Neither the plans nor the name of the architect have survived, but an analysis of the basic town plan, which remained almost unaltered until well into the 19th century, shows clear evidence of creative and far-sighted planning. The official founding Charter was granted by Charles on Mar. 3, 1348—probably the day on which the foundation stone for the city walls was laid. Settlers who wanted to live in the New Town were assigned plots of land and, in return for tax concessions, had to

settlement of Slup, which were planted with orchards, fields and vineyards, was hardly built up at all. It was Charles' wish that the New Town (which at first formed an administrative unit with the Old Town but had to be separated in 1377) should follow the structure of the Old. In the design, the equivalent of the arithmetical street plan—very popular in medieval times—of the Havelské Mesto can be made out.

The generous scale of the plan is most remarkable. The Ječná (Barley Street) is nearly 89 feet (27 meters) wide, and Charles Square with its length of 1,706 feet (520

complete their houses within 18 months.

The New Town was designed to fall into four large sections. The central section is constructed in a geometric plan around Wenceslas Square (the former horse market); the southern portion is centered on Charles Square (formerly the cattle market); and the north is dominated by the former merchant street of Hybernská. The lower area, the valleys and slopes of the former

Left, Charles IV with the imperial regalia. Above, the seal of the Charles University, the oldest university in Central Europe.

meters), an area of nearly 20 acres (8 ha) is the largest square in Europe. However, Charles' building schemes were not confined to the extension of the castle and the New Town. On the left bank of the Vltava a new wall was built and the area of the Mala Strana increased considerably. The Hradčany settlement was also fortified. Part of this fortification is known as the "Wall of Hunger", for the story has it that Charles had this wall built during a time of widespread poverty in order to reduce unemployment. At the same time building began on the Petřín hill. In the Old Town, too, the building boom that had begun in the 1330s continued. It provided the city with the rebuilding of the monastery and church of St. Jacob, the alterations to the churches of St. Aegidius, St. Martin, St. Castullus, St. Gallus, St. Nicholas and also the new building of the Holy Ghost church, to name only a few of the most important churches.

After the tremendous boom time of Charles' rule, Prague, at the end of the 14th century consisted of two castles and four "towns" with an area of 1,976 acres (800 ha) and a population of over 50,000. Within the city area, there were around 100 monaster-

caused in the first instance by the social and national contrasts among the settlers and also between the inhabitants of the Old and New Towns, resulted in a break between the two settlements in 1377. When Charles IV died on Nov. 29, 1378, he left Prague with a lot of building sites, and also with a lot of problems smouldering under the surface.

Wenceslas IV

Charles' son Wenceslas IV succeeded Charles' position in the Empire and in Bohemia. He faced strong political opposition

ies, churches and chapels, several dozen markets and an impressive system of water supplies. Prague had been promoted to an imperial residence, an archdiocese, the seat of a papal legate, and a university town. Because of its great political and cultural importance, Prague was able to counter the economic disadvantages caused by the moving of the trade routes to the Danube by importing and trading in luxury goods.

However, despite intensive effort on Charles' part, the dependence of the city economy on non-economic political factors increased. The uneven social structure,

within the Empire and had to accept a considerable loss of power and authority. Because of this, Prague decreased in importance and kudos. Building works soon slowed down, economic difficulties arose and led to a depression that brought social unrest in its wake. The dissatisfaction of large sections of the population, especially the poorer Czech inhabitants of the New Town, was focused on the rich (mostly foreign) patrician families and particularly on the clergy.

In 14th-century Europe, a general opposition to the luxurious and often not very moral lifestyle of the monasteries was growing. In

Prague this had already come under critical fire, occasionally from members of its own ranks, during the reign of Charles IV. In the late 14th century, Konrad Waldhäuser (died 1369) and Jan Milic Kromeriz (died 1374) were prominent preachers who attacked the luxury and immorality in the monasteries.

In 1391, the Bethlehem Chapel was founded. Its plain exterior alone (the chapel was rebuilt exactly to original plans in the years 1950-53) marks it as totally separate from the criticized conditions. In March 1402, Jan Hus (1369/70-1415) began to preach in this chapel against the seculariza-

tion of the church.

The 45 Articles of the English theologian John Wiclif (1328-1384) had a considerable influence on the reforming work of Jan Hus, who urged a re-awakening of the church, based on a return to the message of the Bible and a lessening of the gap between clergy and laity. Hus' ideas were very popular among the citizens, and even with Wenceslas IV. However, the clergy, afraid of los-

Left, Jan Hus being led to the stake in Constance. Above, In the shadow of his father Charles: Wenceslas IV.

ing power, rejected him decisively.

Jan Hus

In 1398, Hus was appointed to the University as "magister primarius", as Professor of Philosophy. Here he continued to expound his ideas for reform. However, Hus was defeated in a vote by the "nations"—most of the academics condemned his and Wiclif's thinking. In 1409, the dispute reached a crisis. It was no longer purely a theological quarrel, but now had nationalist overtones. This was the year in which Hus managed to obtain the Decree of Kuttenberg (Kutna Hora) from his patron Wenceslas IV. This granted the Bohemian "nation" in the University a majority of votes.

After making strong protests, the German academics moved out of Prague in a body and began an Empire-wide campaign against the University of Prague. At the same time, the clergy of Prague reacted to Hus and his followers with arrests and repressive measures, as did the patrician families, who are greatly disturbed by the social and political applications of Hus' ideas as proposed by Jan Zelivsky.

Escalation of the conflict came on Jul. 30, 1419, when an angry mob led by Zelivsky marched to the Town Hall in the New Town and demanded the release of the arrested Hussites. The consuls refused, and the enraged citizens stormed the Town Hall and began the tradition of defenestrations in Prague when they hurled the consuls and seven other citizens who were defending the Town Hall out of the window. Unrest spread rapidly, and could no longer be controlled even by royal troops. Soon the Hussites occupied the Town Hall and elected their own consuls. Wenceslas made no effort to supress them and, in August, approved the appointment of their consuls.

Perhaps he hoped to tone down the political conflict with this move, but the signal came too late. Wenceslas IV died on Aug. 16, 1419 in Novy Hrádec. On August 17, the Hussites stormed the Carthusian monastery in Ujezd and continued the revolution.

RUDOLF II AND THE HABSBURGS

The winners in the Hussite revolution were, for the most part, the Czech nobility. During the decade from 1430 to 1440 the Catholic city governors were driven out, church property was confiscated, Prague's independence was legally confirmed—triumphs for the Bohemian Estates. Only Czechs of Hussite persuasion were allowed to vote on the Prague city council. Rome agreed to religious freedom with the so-called Compacts of Prague. However, the Catholic church was merely biding its time.

But for now the people of Prague acclaimed their first Hussite king. In 1458, George of Podiebrady was crowned. A Czech and an Utraquist, he was King of Bohemia for 13 years, and Prague, laid waste by the revolution, its economy in ruins, blossomed once more. George had the towers of the Teyn church and the Bridge Tower in the Malá Strana built. He had no time to do more, as Rome incited people against the "heretic king". During the rule of George's successor, the Polish prince Vladislav (1471), the second Defenestration of Prague occurred. Vladislav had let the Catholics back in. They occupied the Old Town Hall, arming themselves against the Utraquists. But then dark plots were made public. During the nights of Sept. 25 and 26, 1483, the populace stormed the Town Hall and threw the spokesman, and the mayor of the Old Town out of the window. In 1484, Catholics and Utraquists made peace once more with the Treaty of Kuttenberg (*Kutna Hora*). The unrest had badly affected King Vladislav, whose residence was in the royal palace in the Old Town. He moved and chose the dilapidated castle as his new residence and renovated it. The masterpiece is the Vladislav Hall, the first hint of the Renaissance style among the Late Gothic prevalent in Prague, and was built by

Preceding pages: view of Prague c. 1493. Left, Van Dyck portrait of General Wallenstein. Above, Prague flourished once again under the rule of Rudolf II.

Benedikt Rieth (1454-1534).

It was the teaching of Martin Luther that split the Utraquists into two camps during the first half of the 16th century. The Old Utraquists were against the German Reformation, the New Utraquists supported it. Catholicism also found support from Vladislav's son Louis, who began his reign in 1516. Following Louis' early death the divided Estates chose a new king in 1526. They made a fatal choice: the new King of Bohemia was the Austrian Archduke

RODOLPHVS II. MAXIMILIANI II FILIVS. NATVS XV KAL. SEXT. CIƆIƆLII. HVNGARIÆ. REX. DEINDE. ET BOHEMIÆ. PATRIS DEFVNCTI OBTINET LOCVM IMPERATOR. CIƆIƆLXXVI. OTTOMANNOS DIVTVRNO BELLO INFIGNITER CASTIGATOS TRIVMPHAT. PACIFICVS OBIT PRIDIE. III. IAN. CIƆIƆCXII.

Ferdinand I of Habsburg (1526-1564). Prague became the most important prop of Viennese rule, and Rome found more and more support for its fight against the Utraquists. The Hussite era was facing its final defeat. Ferdinand soon quarreled with representatives of the Bohemian Estates.

It was Ferdinand I's pleasure palace of Belvedere that finally helped the Renaissance to establish itself in Prague. The Belvedere, on a hill opposite the Hradčany, became the model for the palaces of the nobility. John of Lobkowic, for instance, had his palace built in the latest Italian style.

Later, this building came into the hands of the Schwarzenberg family. It still bears their name today, and is now the Museum of Military History.

In 1555, Ferdinand had the hunting lodge Star Palace on the White Mountain. By this time Prague had surrendered unconditionally to his rule. However, in 1546 open conflict broke out once more. Ferdinand went to war against the Protestant German princes. The Estates in Prague refused to support the King in a war against their co-believers. However, the King returned victorious, with clearly one thought on his mind: revenge! On Jul. 1,

independent, as the Habsburg residence. Many historians compare the Rudolfine era with the glorious reign of Charles IV. Life came back into the city: diplomats, political observers, adventurers, traders from all over the world, craftsmen, professional soldiers, musicians and many artists followed the ruler. While Rudolf indulged his passion for collecting works of art, political and religious conflicts continued to seethe, hardly noticed by those in the castle. Meanwhile, Rudolf hoarded art treasures— paintings and drawings were his particular favorites. His favorite painters were Dürer

1547, Ferdinand's mercenaries swooped down on Prague. Never again would Prague deny him obedience. Ferdinand took all the privileges won by the glorious Hussite revolution away from Prague. The city became a vassal of the Habsburgs. All public property had to be surrendered. And now the way was open for the return to power of the Catholic church in this Protestant country. In 1561, Ferdinand succeeded in appointing a new Bishop of Prague—the post had not been filled since the Hussite revolution.

However, Prague acquired new fame in 1583, although it was no longer politically

and Peter Brueghel the Elder. If he couldn't get hold of the original, he had it copied by Jan Brueghel or Peter the Younger. Rudolf also piled up paintings by other leading artists of the Renaissance: Titian, Leonardo, Michelangelo, Raffael, Bosch and Corregio. The Emperor was also interested in all kinds of curiosities and rare objects. An inventory records: "In the two upper compartments all kinds of strange sea fish, underneath them a bat, a box of four thunder stones (meteorites), two boxes of lodestones, and two iron nails, said to come from Noah's Ark, a stone that grows which was a gift from Herr von Rosenberg, two bullets taken from a Transylvanian mare, a box of mandrake roots, a crocodile in a bag, a monster with two heads..."

Despite his somewhat indiscriminate passion for collecting, Rudolf II made Prague into the "artistic treasure house of Europe". Unfortunately much was lost during the Thirty Years' War. There was little construction in the building industry at this time. The Jesuits began in 1578 with the great St. Savior's Church. The basic construction still adheres to the old Gothic pattern, but windows and sills, reliefs and the

freedoms. In 1611, after a failed attack on Prague, Rudolf had to surrender the crown to his brother Matthias. Barely a year later Rudolf died.

The Thirty Years' War

In all ages there are (or were) places in the world where political observers can feel the pulse of the times beating faster than elsewhere. Prague in the early 17th century must have been such a place. The gulf between the House of Habsburg and the Bohemian nobility, between Catholic and

vaulting are built using a more modern style.

The great church of the Jesuits was the symbol of growing Catholic strength in Prague. The conflict between Rudolf II and his brother, Archduke Matthias, was a long-awaited trump card for the Protestants. Rudolf, under pressure from Matthias, had to make concessions. In 1609, the Estates forced the king to issue the Majestát (Letter of Majesty) guaranteeing religious

Left, Prague's forces fight against the Swedes on the Charles Bridge (1648). Above, the murder of Wallenstein in Eger (1634).

Protestant was a reflection of the political situation throughout Europe. Indeed, the conflict here had a long tradition unparalleled anywhere else. No one should be surprised, then, that Prague was the place over which the storm clouds of war, which had been hovering threateningly over all of Europe, broke first.

"Follow the old Czech custom—throw them out of the window!" a voice from the crowd is supposed to have shouted on May 23, 1618. The representatives of the Bohemian Estates were enraged and stormed the Court Chancellery in the castle.

Protestant churches were burning in the surrounding countryside. In 1617, the oppressor of the Protestants, Ferdinand of Styria, had been crowned King of Bohemia. Count Martinic, Governor Slavata and Secretary Philipp Fabricius fell nearly 55 feet (16.5 meters) into the castle moat, and all of them survived. Prague was in uproar. The Thirty Years' War began and Prague was the center of the revolt.

A year after the Defenestration, the Bohemian Estates got rid of Ferdinand II and made Frederick V, Elector of the Palatinate, King of Bohemia. But Ferdinand II was a

Joseph II. Empereur des Romains

Habsburg Emperor based in Vienna, and he was going to pay them back.

On Nov. 8, 1620 the combined armies of the Emperor and the Catholic League were drawn up on the White Mountain outside Prague, facing the army of the Bohemian Estates. The outcome of the battle was decided within a few hours. The army of the Estates fled behind the walls of Prague, and King Frederick to the Netherlands. Prague was defeated. For weeks enemy troops plundered the city and the damage ran into millions of guilders. A dreadful revenge indeed. The presumed leaders of the revolt

were arrested and executed. Even before the executions were carried out the Emperor had re-instated the Catholic clergy. On June 21, 15 citizens of Prague, 10 members of the nobility and two citizens of other estates were executed in the Old Town Square. The heads of 12 directors were fastened to iron hooks on the the Bridge Tower in the Old Town, as a permanent warning. A cruel penance for the Defenestration. In the same year all non-Catholic clergy were forced to leave Prague. The Emperor had torn up Rudolf's Letter of Majesty with his own hands. In the years that followed many families emigrated.

During the following thirty war-torn years only a few new buildings arose in Prague. One of them was the palace of Albrecht von Wallenstein (or Waldstein) in the Malá Strana, built in the years 1624-1630 on a site that had formerly contained 23 houses, a brickworks and three gardens. The Baroque style became fashionable at that time. Churches that belong to the Baroque era are the church of St. Nicholas in the Malá Strana (1704-1755), the church of St. Nicholas in the Old Town, the church of St. Katherine in the New Town (1737-1741) and the Holy Trinity Church in the New Town (1720). The most famous architect of the Baroque period was Kilian Ignaz Dietzenhofer, who died on Dec. 18, 1754.

Under the rule of Maria Theresia (1740-1780) and Joseph II (1780-1790) religious freedom returned to Prague. Joseph II proclaimed the Edict of Tolerance in 1781. The age of religious wars was at an end. But was freedom of religion the only aim of the Bohemian Estates? We should remember the decade 1430-1450 and the achievements of the Hussite revolution—at that time only Czechs were allowed to sit on the town council. Four hundred years later, the inhabitants of Prague rediscovered their national consciousness.

Above, Joseph II, the son of Maria Theresia. Right, Maria Theresia, Empress of Austria and Queen of Bohemia (1717-1780)

The Defenestration of Prague, on the eve of the Thirty Years' War, was the last flare-up of post-Hussite Czech national consciousness. And it was the Battle of the White Mountain in 1620 that crushed it utterly. Twelve heads hung on the Bridge Tower, gruesome symbols of the destruction of Czech culture.

In the centuries that followed, the Czech language disappeared from the Estates, and the Czechs became a people of peasants, small-time craftsmen and servants. In the 18th century, the upper classes, nobility and bourgeoisie, were German. Language differences often made communication between people and administrators impossible. Maria Theresia commanded that all justices and civil servants should know the "vulgar tongue". Teachers taught Czech once again.

The Czech Language Comeback

The late 18th century was the great period of the theater in Prague. The first Czech performance was held as early as 1771. In 1781, the Nostitz Theater was built by Count Anton Nostitz-Rieneck (today it's the Tyl Theater). When the theater opened, the German upper classes went to see Lessing's *Emilia Galotti* or, in 1787, applauded Mozart's *Don Giovanni*. While this was happening, the Czechs struggled to put on matinées in the Great House. In 1785, they gained permission to do so for a short time, but soon Czech performances in the Nostitz Theater were banned once more.

The Czech players moved to the Bouda (booth) in the Horse Market. This was a little wooden theater. It was to be another 60 years before the Czech nation was to appear on the stage of history, but the Czech language could no longer be suppressed. Czech books and newspapers returned, and Czech was taught at the University once more.

In 1833, the Englishman Edward Thomas began the production of steam engines in Karlin. The rapid economic development of the Industrial Revolution created an indus-trial proletariat in and around Prague and the number of Czechs in the population of Prague increased. Tension increased between Germans and Czechs. However, at first both had a common enemy: the all-powerful Viennese State Chancellor Metternich. The year of revolution, 1848, was the last time that Germans and Czechs together manned the barricades for a common cause.

Even then the goals were not really common to both parties, for the Czechs were no longer really interested in Bohemian, but in Czech freedom. In February 1848, there was revolution in Paris, and Metternich resigned in Vienna on March 13. There was rejoicing in Prague. But the rejoicing was divided—the Germans wanted to accept the invitation of the revolutionary Frankfurt Parliament, the Czechs wanted their own state as part of a federal Austrian Empire.

Violence at Slavic Congress

The Slavic Congress met on Jun. 2, 1848 in the museum building of Prague. One of its demands was for equal rights for all nationalities. The leader of the movement was the Czech František Palacky: "Either we achieve a situation where we can say with pride: 'I am a Slav', or we shall stop being Slavs". The Congress came to a bad end. After a Slavic mass the Prague militia fired into the crowd, and the nationalist movement was crushed in bloody fighting in the streets and on the barricades.

From 1849 on the citizens' revolutions in all the nations of the Austrian Empire came under the rod of absolutism. The Czech language was even forced back out of the civil service departments. But after the war of France and Italy against the Austrian monarchy, absolutism was at an end. On Mar. 5, 1860 an "extended imperial council" was called. Representatives from different lands

Right, Tomas Masaryk, a leading Slavic nationalist, chief founder and first president of the Czechoslovakia republic (1918).

sat with the councillors appointed by the Emperor. In 1861, a Czech could become mayor of Prague, and yet on a national level the Emperor favored the German-speaking Bohemians. The Czechs remained loyal to the Emperor in the struggle with Prussia in 1866 for supremacy in Germany, and yet they were not rewarded. There was no Czech equality, never mind autonomy.

"Libussa" in the National Theater

The Czechs could only win minor battles— for instance, they finally got their long-

gulf existed between Germans and Czechs. Austria and Hungary mobilized after the shots were fired at Sarajevo on Jun. 28, 1914, but meanwhile in Vienna, German and Czech Bohemian parties were discussing the reform of the state laws, state electoral reforms giving equal voting rights to everyone and the regulation of the use of the two languages. At that time hardly anyone— not even the optimists—gave serious consideration to a Czechoslovak Republic, which actually came into existence five years later. At that time, just before World War I, the Czechs were still fighting the Ger-

awaited Czech national theater in Prague. Even today it is still a symbol of Czech national sentiment. It was opened on Jun. 15, 1881 with a special performance of Smetana's *Libussa*. Unfortunately the theater burned down on August 12 of the same year, but it was quickly built up once more. In 1876, work began on the Rudolfinum, and from 1893 on the Bohemian National Museum was to be found in the Horse Market, today Wenceslas Square. The university was split in 1882—there was now a Czech and a German university.

By the time World War I broke out, a deep

mans for equal rights. Five years later they were to take over the government of their own state.

The Czechoslovak Republic was the work of the exiles Tomas Masaryk, Edvard Beneš and Milan Stefanik. During World War I they had made it very clear to the Allies that a strong Slavic bloc between the German Empire and Austria was necessary.

Above, shop in the Jewish quarter, before 1900. Right, Edvard Beneš, state president of the first republic and again after 1945.

BOHEMIA AND MORAVIA—
THE REICH PROTECTORATE

Rejoicing in the Streets

On Oct. 28, 1918 the Czechoslovak state was declared in Prague. But the borders were still not confirmed: the Sudeten Germans wanted to join with German-speaking Austria, the inclusion of Slovakia in the Czech republic had not yet been decided, and Poland was claiming the coal mines in the former dukedom of Teschen. On November 14, Tomas Masaryk was elected President of the Republic and was wel-

autonomy for Slovakia". The incorporation of the Sudetenland had even worse consequences. Later, it gave Hitler a welcome excuse to liquidate Czechoslovakia. Following the formula of the American President Wilson regarding the right of peoples to self-determination, representatives of the Sudeten Germans had declared an "autonomous province of the state of German Austria" on Oct. 28, 1918. Towards the end of that year, Czech troops retaliated by occupying German-settled areas. The Peace Conference

comed back by the enthusiastic people of Prague after four long years of exile. Foreign Minister Edvard Beneš—Masaryk's successor in presidential office—made skilful use of the last weeks of the war and the time after Germany's surrender. The uneasy powers of the Entente had no clear idea of how Europe should look once peace was declared. In 1918, the Czech foreign minister managed to obtain the incorporation of Slovakia into the Czech state, against Hungarian opposition and the opposition of the populace, who also wanted autonomy. Continual internal unrest was guaranteed by demands for "national

decided in favor of Czechoslovakia. Without a plebiscite, the German areas went to Czechoslovakia. Those parts of Teschen which were incorporated were not all in favor, either. After the Treaty of Versailles, opposition arose all over the country.

The Sudeten Germans saw themselves as an oppressed minority, disadvantaged by language rulings, land reform, and the unfavorable position of the German education system and industry. The economic effects of "Black Friday" on Oct. 4, 1929 strengthened radical opinion. Around two-thirds of the 920,000 unemployed in the winter of

1932/33 were Germans. In 1933, the gymnastics teacher Konrad Henlein founded the "Sudetendeutsche Heimatfront" (SHF = Sudeten German Home Party). In 1935 the SHF, now the "Sudetendeutsche Partei" (SdP = Sudeten German Party) took part in the elections and soon became the voice of the people in the German-settled areas. It did not take long for Henlein to make contact with Hitler. After a few years he and his party became puppets of their mighty patron. This was why the SdP's demands became more

to pressure from Hitler. On Sept. 30, 1938 Chamberlain, Daladier, Mussolini and Hitler signed the Munich Agreement. The Sudeten lands now belonged to Germany, Czechoslovakia had been sacrificed. On October 22, Beneš went into exile in England, and on Mar. 14 and 15, 1939 the fate of Czechoslovakia was sealed. Hitler declared a "sovereign" Slovakian vassal state and established the Protectorate of Bohemia and Moravia. German troops marched into Prague on March 15 without meeting any

and more radical over the years, and why their goals shifted from equal rights, more and more openly, to inclusion in the German Reich. In 1937, Henlein drew the conclusion "that today even the broad mass of Sudeten Germans no longer believe in equal rights with the Czech people in a Czech state."

In the autumn of 1938 the time was ripe. Afraid of war, France and Germany gave in

form of resistance. The time of worst oppression began. The chief of the SD (Sicherheitsdienst = security forces), Reinhard Heydrich, attacked the Czech intelligentsia and the middle classes. After Heydrich's assassination, this process was continued by General Kurt Daluege. The Czechs were defined as second class persons under Article 2 of the Treaty of the Protectorate. Czech universities were closed. Academics were not allowed to work in their professions, thousands were arrested and imprisoned in the concentration camps of Dachau and Oranienburg.

Preceding pages: St. Wenceslas Square, before 1900. Left, Sudeten German women acclaim Hitler. Above, crossing the border to Czechoslovakia.

FROM PEOPLE'S DEMOCRACY
TO SOCIALIST REPUBLIC

The Situation in 1945

In post-1945 Czechoslovakia, the path to socialism had been smoothed by history as in no other country. Before World War II, the country was one of Europe's highly developed nations, with modern light and heavy industry, and efficient agriculture. Above all, it possessed a self-confident, highly qualified proletariat and a class of educated people, intellectuals and artists, who were, if not actually members of the Communist Party of the time, were nonetheless mainly left or liberal-left in their views. The relationship of the Czechs to the Soviet Union and to socialism was positive. Following the rebirth of Czech national consciousness in the early 19th century, the Czechs viewed the Russians as a kind of Slavic older brother.

Apart from two exiled groups, one in England and another group in Russia, the Czechs did not fight against Nazi Germany in World War II. The number of victims that they mourned in 1945 was small compared with those from Poland, Russia or the Ukraine. However, no other people in Europe had lost such a high percentage of their intellectuals and artists under Nazi rule. In the years from 1939 to 1945, the Nazis in the Protectorate of Bohemia and Moravia systematically exterminated not the people, not the workers, not even the technologically qualified, but the academic and creative elite of the nation.

When World War II ended in May 1945, the Czechs enthusiastically greeted the soldiers of the Red Army (who had occupied much of Czechoslovakia) as liberators and Slavic brothers.

In February 1948, the communists gained control of Czechoslovakia in a bloodless coup. At that time they could still count on

Left, announcement to the crowds in the Old Town Square. Above, the sudden end of the "Prague Spring"—the dream of a humane socialism is over.

the support of the majority of Czechoslovak people. However, bitter disillusionment was just around the corner.

The Stalin Years

The name of the disillusionment was Stalin, who mercilessly forced the Czechoslovaks to accept his particular version (or perversion) of communism. The awakening from the Bohemian dream of social justice was cruel indeed. In a period of eight years,

from 1948 to 1956, the Czechoslovak Communist Party, by now Stalinist in orientation, succeeded in destroying any idea of a specifically Czechoslovak way to socialism thoroughly and possibly for years to come. The rule of the Stalinists had very negative consequences for Czechs and Slovaks. In the late 50s and early 60s skepticism and cynicism spread throughout Czechoslovakia. Hardly anyone still believed in Marxism or Leninism, or even in the idea of just and fair socialism. Also, Czechoslovakia was at rock bottom, economically speaking, by 1963. The events, five years later, which came to be called the "Prague Spring", had already begun in 1963 with the failure of the so-called Five Year Plan and the near collapse of the whole economy of the country.

Literature and Politics

In those years, called by some as the "golden five years" before the Prague Spring of 1968, a literature independent of the previously all-powerful Party censorship machine was developed in Czechoslovakia. Literature is of particular importance because, ever since the re-birth of the Czech

national consciousness in the early 19th century, Czech writers have had a vital role to play. When the crisis of Czechoslovak society and socialist beliefs came, in the early 60s, it was once again authors such as the future Nobel Prize winner for Literature, Jaroslav Seifert, the lyric poets Vladimír Holan, František Halas and others, who replaced the helpless functionaries from Party headquarters as political and moral institutions. By 1965, at the latest, the Czechs and Slovaks had realized that Czechoslovak socialism could not continue in this manner, and, searching for something

to cling to, they rediscovered their poets and authors. "Socialism with a human face" was a program that was developed not by the Communist Party and not by the exhausted and insecure ideologies, but by the poets of the 1968 Prague Spring.

End of a Dream

"Socialism with a human face" was the demand made in the spring of 1968 by the writers and by Dubček's reforming communists. Behind this demand lay a bitter admission and an undertone of despair: up until the spring of 1968 socialism in Czechoslovakia had obviously not shown a very human face.

The dream of an efficient, just and happy socialist society came to a final end in Czechoslovakia with the collapse of the Prague Spring in 1968. The Bohemian ill fortune that had dogged so many of their predecessors, had also dragged down the heroes of the Prague Spring.

Bohemian history is full of absurdities. In 1968, the Soviet Union marched in with soldiers from five socialist countries and destroyed the Czechoslovak hope of socialism with a human face. Today, 20 years later, the Soviet Union is searching for the hope destroyed in Prague in August 1968, and finds it in the political programs of the Prague reformers, who, back at home, are still considered to be counter-revolutionaries and reactionaries.

A Look Forward

Over 100 years ago socialism was the great dream of the people of Bohemia and Moravia, the industrial lands of the old imperial monarchy. Today it has come to a dead end. Today, it is Gorbachev's reform policies, a kind of Prague Spring in Moscow, which have sown the seeds of new hope.

Above, State president Husak. Right, Alexander Dubček in 1968. His vision of "socialism with a human face" is once again reflected in current trends of thought.

EVERYDAY LIFE IN PRAGUE

Everyday life in Prague is full of problems—political, social, logistical, and, last but not least, problems of human relationships. We can only hint at these problems here. At first glance, the problems seem terribly complicated and can only be resolved by means of bribery. Please don't be shocked by the word "bribery"—in Eastern European countries, bribery is one of the most efficient methods of not only making social relationships possible, but also of humanizing them in the warmest and

really need it. But the bottle as a means of bribery signals that he is facing a human being who thinks exactly in the same way as he does, i.e. not according to strict ideological guidelines but as a pragmatic individualist. You bribe the dentist, and he agrees to be bribed by you; this arrangement brings about human contact—firmer, more enduring and more profitable for both parties than love and marriage, or even many years of membership of the Communist Party! You open not only your mouth, but

most attractive way.

Let us assume that you are an inhabitant of Prague and, at 10 a.m. on a Tuesday, you get a terrible toothache. If you live in the district of Prague 7, then you have to go to one of the dentists responsible for Prague 7. But, as the highest responsible authority has made no plans for toothache on Tuesdays, the dentists are actually not allowed to treat you.

For these and similar cases, which the Party and State Five Year Plan hasn't provided for, a citizen of Prague keeps a bottle of good (preferably Western) spirits. The dentist who accepts the bottle doesn't

also your heart to the dentist, and he opens his to you. For you soon find that you can help and support each other. The dentist is building a dacha and needs heaters, and the socialist industry can't deliver them until after the next Five Year Plan. However, you have a friend who can "organize" heaters for you, because you managed to obtain a place in High School for his very talented daughter (who has, alas, a miserable record of approved political activity), or because you gave the Comrade Principal's wife a birthday present of two dozen towels, which you got from your aunt, who runs a textile shop. Gradually, this process leads to the development of new communities of interest among the citizens, who help each other out and organize their lives, their material and even their spiritual needs around the "unofficial" economy, i.e. outside the structure of party and state planning.

Members of the Communist Party are welcome in these self-help groups, because they have access to and influence in the official departments where decisions are made that affect the fate of members of the

then on, the father of the exiled traitor was suddenly transformed into a good comrade, his youngest son was permitted to study at the university, and he himself could even go on tourist trips to Western countries.

Another excellent example is provided by Comrade Professor Dr. D.H., a great Marxist theoretician and philosopher. The windscreen wipers on his Soviet-made car were broken. Professor Dr. D. H. drove the car to a garage and demanded new wipers to be fixed. "Haven't got any, come back in a

groups and the fates of their children. To give an example: a member of the Communist Party wanted to give his son a Western car as a present. A member of his self-help group has an eldest son living in exile in the West. This son bought a used Mercedes in West Germany, and arranged for it to be driven to Prague. Everything went on smoothly. From

month!" growled the Comrade Manager of the garage.

"But the Party Resolution of 1976 stated that the delays in the supply of spare parts were to be eliminated by 1986!" replied Professor Dr. D. H.

"True. We did have windscreen wipers in July 1986."

"And why don't you have any now?"

"Because we already achieved our goal in 1986," said the manager of the garage, took a swig from his beer bottle and paid no further attention to the totally confused Professor Dr. D. H.

Preceding pages: Metro station Malostranská. Left, queues are part of everyday life in Prague. Above, after the wedding ceremony in the Old Town Hall.

CZECHS AND SLOVAKS

Neither the Czechs nor the Slovaks had considered living together in one state until the end of World War I. The Czechoslovak National Council was formed in 1916 by Czech and Slovak exiles in the USA, but not until the Habsburg monarchy was unquestionably collapsing did the Slovaks in America accept the ideas of the leader of the Czechs in exile at that time (later, President of Czechoslovakia), Professor T. G. Masaryk, and join in signing the so-called "Pittsburg Convention" on May 31, 1918. In

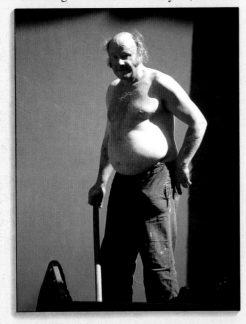

the future republic of Czechoslovakia, Czechs and Slovaks were to have equal individual rights.

Up until 1918 both Czechs and Slovaks lived under the Habsburg monarchy, though Slovakia was counted as part of Hungary. The "Hungarianisation" of Slovakia was carried out with such thoroughness and brutality that by the beginning of the 18th century the Slovak language had ceased to exist. Under the influence of the rebirth of Czech nationalism the Slovaks rediscovered their "lost" language and developed it further. After 1918, the Czechs

and the Prague government tried hard to build up the Slovak education system and to help the Slovaks build up their own intelligentsia. There were Czechs working in Slovak schools and colleges, in administration, in the judicial system and in other departments. Of course, this led to misunderstandings—the Slovaks felt patronized and later even oppressed by the Czechs. However, the Czechs also contributed to this misunderstanding. The "Pittsburg Convention" of 1918 stated that the Slovak people were an equal partner in the new republic, but it was soon forgotten in Prague, and many Czechs looked on Slovakia as their colony.

Discontent in Slovakia found its expression in the program of the Slovak separatists, the Slovak Populist Party. In March 1939, the rest of Czechoslovakia was occupied by Hitler, and the Slovak separatists felt that their moment in history had come. They split off from the republic and, under Hitler's "protection", formed an "independent" fascist Slovak state, allied to Nazi Germany.

By the end of August 1944, the Soviet army had reached the northern borders of the Slovak state. Slovak patriots, together with the communists, organized a popular revolution, which was crushed by the Germans but was of historic importance to the Slovaks. The Czechs and Moravians did not rebel against the Nazis until Germany had already surrendered.

The differences continue today. The Slovaks did not take much part in the Prague Spring of 1968, and the following so-called clean-up of public life claimed fewer Slovak than Bohemian or Moravian victims. That is why many Czechs now sigh "The best job of all nowadays is to be a Slovak in Prague."

Preceding pages: Spartakiad games; view of the Hradčany; young hopeful for the ice hockey team. Above, many Prague apartments are heated by coal. Right, group in traditional dress in front of an inn.

THE FIVE TOWNS OF PRAGUE

Today you can no longer talk about the Five Towns, for Prague has now been divided into ten districts. However, most people are interested only in the five historic towns: Hradčany and the Old Town, Malá Strana and the New Town, as well as the Prague Ghetto, somewhat euphemistically known as "Josephstown" (Josefov). At the beginning of the 19th century, some 80,000 people lived in the city. Gradually more districts were added, and the population grew steadily. Vyšehrad, Holesovice and Bubenec brought the population of the city to around 200,000 by 1900. On Jan. 1, 1922, the area of Greater Prague had a population of 676,000. After World War I, the city grew in leaps and bounds. Its area tripled to a size of 193 sq miles (550 sq km). New suburbs such as Severní mesto (North Town) with a population of 80,000 and Jižní mesto (South Town) with a population of 100,000 were built. The south-western suburb of Jihozápadní mesto, intended for 130,000 inhabitants, is still being built. Today about 1.3 million people live in Prague, and about 10 percent of Czech industry is based there.

City administration was not unified until the rule of Emperor Joseph II. The separate town halls are reminders of previous autonomy. Even today, the individual districts of the city each have a different atmosphere and a different feel. A visit to Hradčany castle and its surroundings, such as the Loreto church or the Strahov monastery, has a distinct atmosphere of its own. The Malá Strana and the island of Kampa with its ostentatious palaces, built in the shadow of the rulers of Hradčany castle, is another, quite separate tour. Its wine bars, the lovely gardens of the Petřín hill, the view of the Vltava and the Charles Bridge by night—they all have a romantic fascination for many visitors to Prague.

In earlier times, the inhabitants of the narrow and dirty Old Town and the Jewish quarter must have felt quite envious when they looked across to the other bank of the Vltava. If you take a look from the beautifully restored Old Town Square at the narrow alleys and courtyards surrounding it, you will get a brief glimpse of old Prague, the Prague of Franz Kafka. Today the Pařížská is a splendid street in which both Dior and Lufthansa have established themselves. But just go a few blocks further on.

The generous, well-planned and far-sighted designs of Charles IV and his architects are apparent when you walk through the New Town. This was planning well ahead of its time. The broad open spaces such as Charles Square or Wenceslas Square are products of a time when no-one had yet dreamed of cars or trams.

Preceding pages: shop window displays; Café Slavia; Charles Bridge with swans. Left, Letna Hill offers an amazing view of the Vltava and the bridges of Prague.

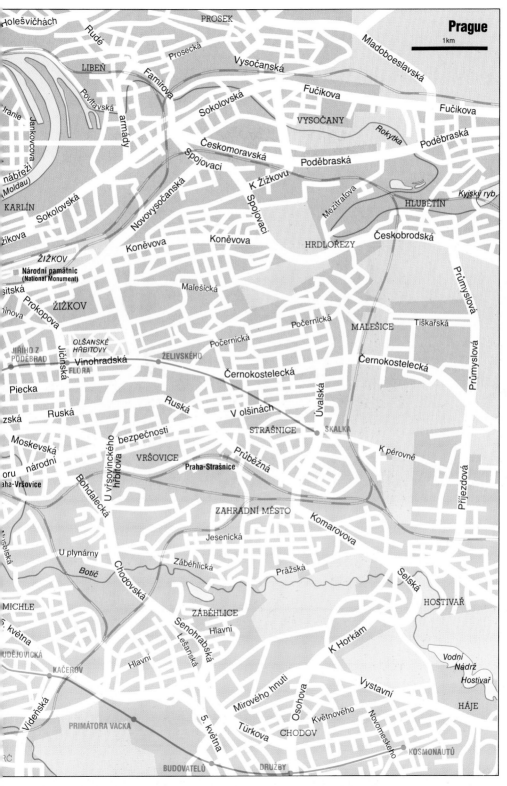

Hradcany Castle

The silhouette of the castle is perhaps the best-known view of Prague. With the advantage of its exposed position, the castle dominates the skyline of the left bank of the Vltava. Especially when floodlit at night, the broad front with the cathedral in the background is most impressive. The historical importance of this royal residence matches its imposing appearance. Its history is not only tied up with that of the city, but also with the history of the first independent Czech state and its future fate. A thousand years ago the fate of the country was decided here, and this tradition has continued, with few interruptions, up to the present day. The castle is the seat of the President of the Republic, and still a center of political power.

The building of the castle dates from the same period as the first historically documented prince of the Přemyslid dynasty. Prince Bořivoj built what was at first a wooden fort on the site of a pagan place of worship. It became the seat of the dynasty and secured the crossroads of important European trade routes which met at the ford of the Vltava. At the same time, Bořivoj built the first church on the hill as a sign of progressive Christianization. In 973, when the bishopric of Prague was founded, the castle also became the bishop's seat.

After the turn of the millennium a Romanesque castle was gradually built up, with a princely (later royal) palace, a bishop's palace, several churches, two monasteries and massive fortifications. Every subsequent period has added its contribution to the complex development of the castle, of which you can get a rough idea when you visit. The castle as we see it today is mostly due to Empress Maria Theresia. In the second half of the 18th century, she commissioned the Viennese court architect Nicolo Pacassi to give the various buildings a unified, Neo-classical facade and extensions. By this means the individual character of the castle has been transformed into something more like a massive palace.

Hradčany Square

Before you start on your tour of the castle, take a look at the Hradčany Square (*Hradčanské nám.*) A few interesting palaces have been built since the destructive fire of 1541, which destroyed all of Hradčany and much of the Malá Strana lying below. Right next door to the castle you can see the Rococo facade of the **Archbishop's Palace**, which is only open to visitors once a year on Maundy Thursday. The Renaissance **Palais Schwarzenberg** on the opposite side of the square has

Preceding pages: at night, the Hradčany castle is floodlit. Left, guard of honor at the entrance to the First courtyard.

painted sgraffito decoration which follows Italian models. In here is the **Museum of Military History** (*Vojenské muzeum*) with its unparalleled collection of weapons, uniforms, medals, flags and battle plans from all over Europe.

The even proportions of its front facade draw attention to the Early Baroque **Palais Toscana**, which closes the square in the west. Where the Kanovnická ul. comes in you will see the Renaissance **Palais Martinic**. When it was restored, sgraffito portraying biblical and classical scenes were discovered.

To the left of the Archbishop's Palace a little alley leads off to the hidden **Palais Sternberg**, which keeps its Baroque splendor inside. This is the main building of the National Gallery (*Národní galerie*), which houses a first-class collection of European art.

On its south side, Hradčany Square opens up to the ramp leading up to the castle, from which you can get a superb view (as you can from the terrace of the *Café Kajetánka*). On weekends in the summer months you can get into the **Castle Gardens** through the entrance beside the **New Castle Steps.**

First and Second Courtyards

The main entrance to the castle complex is the **First Courtyard**, which opens onto Hradčany Square. You enter this so-called **Ceremonial Courtyard** through a gate in an immense ornamental wrought iron fence. The guard of honor is posted in front of the statues, copies of the *Battling Titans* by Ignatius Platzer the Elder. The guard is changed every hour on the hour, a ceremony that always draws a small crowd of onlookers.

This is the most recent of the courtyards and was built on the site of

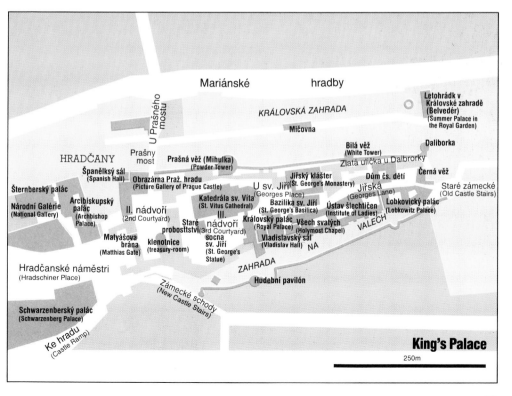

the western castle moat during the alterations of Maria Theresia's reign. Only the Matthias Gate is considerably older, as it is the oldest Baroque building in Hradčany Castle. It originally stood separate, like a triumphal arch, between the bridges that led over the moats. During the rebuilding it was elegantly integrated into the new section as a relief. Since then, the **Matthias Gate** has been the entrance to the Second Courtyard. To the right in the arch of the gate a staircase leads off; this is the official entrance to the public rooms of the President, which are only occasionally open to the public.

The **Second Courtyard** has a somewhat plain appearance. Once there, take a look first at the **Holy Cross Chapel** (*Kaple sv. Křiže*). The most valuable pieces of the **Cathedral Treasure** are kept in this former court chapel with its magnificently decorated interior. They include a collection of valuable reliquaries, liturgical objects and interesting historical mementos. This fascinating collection originated in the days of Prince Wenceslas, but its core goes back to Charles IV. A great pragmatist in political matters, the Emperor was at the same time an impassioned collector of holy relics.

The symmetrical, closed impression given by the Second Courtyard also dates from Maria Theresia's innovations. However, behind this uniform facade lies a conglomeration of buildings which has grown up gradually over the centuries. Each has its own complicated history. In the right passage to the Third Courtyard you can see some remains of the Romanesque castle fortifications.

The remains of an even older building, the church of St. Mary dating from the 9th century, were discovered in the **Castle Gallery**. Access to the

You enter the First Courtyard through the gate dominated by the **Battling Titans.**

gallery is from the passage in the north wing. Here you can see a collection largely put together by art lover Emperor Rudolf II. This unworldly Emperor has gone down in history as an eccentric because of his esoteric way of life, yet he was a great patron of the arts and sciences and collected many art treasures, as well as countless curiosities. His collection was one of the most notable in the Europe of his times. When the imperial residence moved to Vienna, a great part of the collection went with it.

Still more fell to the Swedes as loot during the Thirty Years' War. Yet another valuable collection was built up, still in the 16th century, from the remains, but much of it was taken to Vienna or sold to Dresden. What was left was auctioned off, and was thought for a long time to be totally lost. Only in recent years were pictures discovered during rebuilding work. These were restored and then identified as original paintings which had been believed lost. This small but valuable collection contains 70 paintings (among others, works by Hans von Aachen, Titan, Tinteretto, Veronese, Rubens, M. B. Braun, Adriaen de Vries and the Bohemian Baroque artists J. Kupecky and J. P. Brandl).

St. Vitus' Cathedral

Once you walk through the passage and onto the **Third Courtyard** you can hardly avoid stopping and letting your eye follow the daring vertical lines of St. Vitus' Cathedral (*Chrám sv. Vía*). Only a few steps to the towering north portal is located the biggest church in Prague. It is also the metropolitan church of the Archdiocese of Prague, the royal and imperial burial church and the place where the royal regalia are proudly kept.

The unique 600-year history of the building of the cathedral began when the archbishopric was founded in 1344. Ambitious as ever, Charles IV used this opportunity to begin the building of a cathedral which was intended to be among the most important works of the 14th century Gothic that was spreading from France. To this end Charles employed the French architect Matthias of Arras, who was working in Avignon (a papal city during those years). After eight years the work was taken over by Peter Parler, who influenced all later Gothic architecture in Prague. After his death his sons continued the work, until it was interrupted in the first half of the 15th century by the Hussite wars. In this period the choir with its chapels and part of the south tower were completed. Only a few small alterations were made during the years that followed. For instance, the tower was given a Renaissance top some time after 1560. A good two hundred years later this was replaced by a Baroque roof. The

The banner of the president flies over Prague Castle.

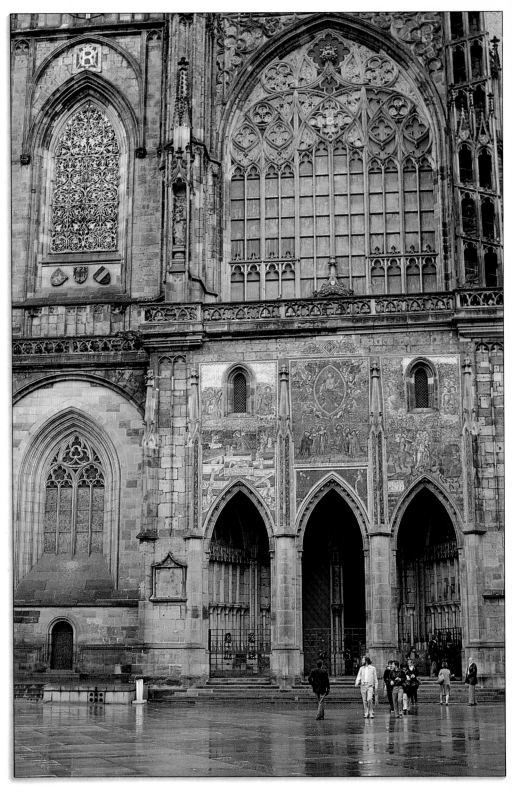

difficult task of completing the cathedral was not attempted until the early 60s of the last century, when a Czech patriotic association took it up. Following old plans and consulting famous Czech artists, they completed the building in 1929.

Before entering the cathedral through the western portal, take a look at its exterior, which dates from the last few years of the completion process. A notable feature is the **Rose Window**, more than 30 feet (10 meters) in diameter, which portrays the creation of the world. On either side of the window are portraits of the cathedral architects. The towers are decorated with the statues of 14 saints. In the center of the bronze gates the history of the building has been portrayed, on the sides you can see the legends of St. Adalbert and St. Wenceslas.

In the splendid interior of the cathedral the most notable features are the stained glass windows and the triforium, a walkway above the pillars with a gallery of portrait busts. Leading Czech artists took part in creating the windows, among them Max Svabinsky, who was responsible for the window in the first chapel on the right, the mosaic on the west wall and the great window above the south portal. The window on the third chapel on the left was designed by Alfons Mucha, who is perhaps best known outside Czechoslovakia for his art nouveau posters for Sarah Bernhardt. If you want to study all 21 chapels (each one contains several works of art), you should join a guided tour. Here we can only draw your attention to the most important sights.

St. Wenceslas' Chapel

The main attraction is bound to be the **Chapel of St. Wenceslas** which protrudes into the transept. It was built by Peter Parler on the site of a Romanesque rotunda of the 10th century, in which the national saint Wenceslas was interred. In keeping with the importance of the Wenceslas cult, the saint's sacred place is exceptionally splendid in its ornamentation. The frescos on the walls, which are decorated with semi-precious stones and gold bezants, portray (in the upper half) Christ's passion and (in the lower) the story of St. Wenceslas. A little door leads to the **Treasure Chamber** directly above the chapel. Here the Bohemian royal regalia are kept, behind seven locks, the seven keys of which are kept by seven separate institutions. However, the precious jewels are only put on display on special occasions.

The three central chapels of the choir, behind the main altar, contain the Gothic tombs of the princes and kings of the Přemyslid dynasty. They are the work of Peter Parler's masons. In the choir itself, on the one side, is a kneeling bronze statue of Cardinal von Schwarzenberg (the work of the leading Czech sculptor J.U. Myslbek, 1848-1922), and on the other side is the massive silver tomb of St. John Nepomuk, designed by the notable Baroque architect Johann Emanuel Fischer von Erlach. Also remarkable are the wooden reliefs in the choir, masterpieces of Baroque woodcarving.

Opposite the tomb of Count Schlick, designed by Matthias B. Braun, a staircase leads down to the **Royal Crypt**. Here you can see the remains of the walls of two Romanesque churches, and also the sarcophagi of Charles IV, his children and his four wives, George of Podiebrady and others. The Emperor Rudolf II lies in a Renaissance pewter coffin. Above the Royal Crypt—just in front of the Neo-gothic High Altar—is the imperial tomb of the Habsburgs, built of white marble for Ferdinand I, his wife Anna and their son Maximilian.

The organ loft originally marked the end of the choir on the west side. Once the Neo-gothic part of the cathedral was completed, it was moved to its present position.

Third Courtyard

In order to see the rest of the interesting sights, you have to walk around the former **Old Chapter House**, (nowadays the House of Culture) which is pressed up against the side of the cathedral. Of special interest to art historians is the equestrian statue of St. George which stands prominently in the courtyard. It is, however, a copy of a Gothic sculpture. The original is in the St. George monastery and is evidence of the highly developed art of 14th century metal casting. The flat-roofed shelters next to the cathedral are to protect archaeological discoveries made in the lower levels of the castle courtyard.

From here you can get an impressive view of the complex system of buttresses and the south facade of the cathedral, which is dominated by the 300 ft (nearly 100 meters) tower. Its stylistically unusual top gives it an individual appearance. The gilded window grille, the letter "R" and the two clocks (the upper shows the hours, the lower the quarter hours) date from the time of Rudolf II. Unfortunately it is no longer possible to climb the tower. It contains four Renaissance bells, among them the biggest bell in Bohemia, which weighs 18 tonnes.

The triple-arched anteroom of the portal that leads into the transept (also known as the **Golden Gate**) has an exterior mosaic depicting *The Day of Judgment*. It was created by Italian artists around the year 1370. The anteroom is fitted with a grille depicting the individual months.

The covered staircase in the left-hand corner leads to the castle gardens,

Detail of the Rose Window of St. Vitus' Cathedral. It portrays the Creation.

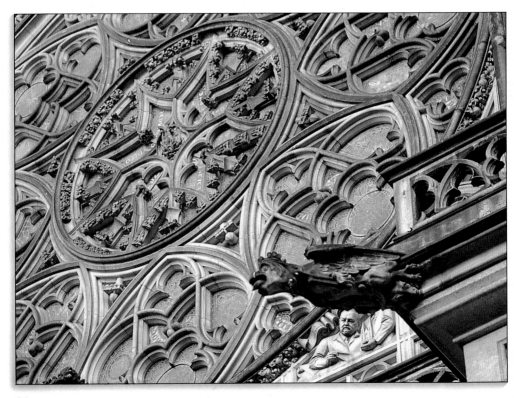

mentioned at the beginning of this chapter. These lead to the former **Royal Palace** (*Královsky palác*) which should definitely be seen. It should no longer be a surprise that this complex was also built by many different generations of rulers. New storeys of the palace were layered one above the other on top of the oldest walls, which now lie deep under the level of the courtyard. Now you can, in a manner of speaking, literally gain deeper and deeper insights into the past.

A Royal Tour

A few years ago the roof of St. Vitus' Cathedral was still open to the public.

Go past a fountain featuring an eagle, and from the courtyard you will be able to go up the staircase leading to the anteroom. From here, you can start your tour of this palace, which up until the 16th century was the residence of the rulers. The first three rooms to the left of the entrance constitute the **Green Chamber**, a former law court and audience hall (it has a ceiling fresco, *The Judgement of Solomon*). Further along is the so-called Vladislav Bedchamber and the Land Records Depository. The Land Records were books in which not only the details of property ownership but also the decisions of the Estates and of the law courts were recorded.

Leave the anteroom and go on to the **Hall of Vladislav**, named after King Vladislav II. This unique, most imposing Late Gothic throne room was built by architect, Benedikt Rieth between the years of 1493 and 1502. Numerous coronations and tournments took place under the 43 ft (13 meter) high pillars.

To the right of the entrance of this hall another wing of the building is joined. Continue on the same level and you will come to the **Bohemian Chancellery**. In the first room is an

attractively clear model which shows the appearance of the castle in the 18th century and compares it to today. Go through a Renaissance portal and you will enter the actual office of the former imperial governor. This room became famous as the site of the so-called "Second Defenestration of Prague", which marked the beginning not only of the Bohemian rebellion but also of the Thirty Years' War. On May 23, 1618 two Catholic governors and a secretary were thrown out of the left-hand window because they had broken the terms of Rudolf II's "Letter of Majesty". A few years previously, the emperor had guaranteed the Bohemian nobility freedom of religion with this decree. Two obelisks in the garden mark the spots where the two honorable gentlemen are supposed to have landed. No such honor was accorded on the secretary. All three survived the defenestration, for they allegedly fell into a dunghill in the castle moat.

Go up a spiral staircase and you will come to the **Imperial Court Chancellery** which lies above the Bohemian chancellery. Under Rudolf II's rule, the whole Holy Roman Empire was ruled from here.

Saints and Fortifications

Under the three Renaissance windows on the narrow wall of the Vladislav Hall a staircase leads off to the **Chapel of All Saints**, which contains three remarkable works of art: the *Triptych of the Angels* by Hans von Aachen, the painting of *All Saints* on the high altar by V.U. Reiner and, in the choir, a cycle of paintings by Dittmann. The latter portray 12 scenes from the life of St. Procopius, who is buried in the chapel.

The next room leading off from the Vladislav Hall is the **Council**

St. George's Fountain in the Second Courtyard.

Chamber, in which the Bohemian Estates and the highest law court assembled. The royal throne and the furnishings are 19th century. To the left of the throne is the tribunal of the chief court recorder, built in a Renaissance style. On the walls are portraits of the Habsburg rulers.

The last room open to the public in this wing is the **New Land Records Office**, with the heraldic emblems of the Land Records officials on the ceiling and along the walls.

The **Riders' Staircase**, which could be used to reach the Vladislav Hall on horseback, leads out of the most recent part of the palace. If the lower storeys are accessible you can continue your tour to the left, going down into the Early Gothic levels of the palace. The lowest level is the Romanesque palace which contains the remains of fortifications partly dating from as early as the end of the 9th century.

Leave the Royal Palace and go out into St. George's Square (*Nám. U sv. Jiří*. The Baroque facade opposite the end of the choir of St. Vitus' Cathedral belongs to the Basilica of St. George (*Bazilika sv. Jiří*). This is the oldest church still extant on the site of the castle. Together with the adjoining monastery, it formed the center of the castle complex in the early Middle Ages. It was founded in the early 10th century and by the 12th century the Romanesque church had attained its present form. Subsequent changes were removed during the rebuilding undertaken at the beginning of this century. Only the Baroque facade and the southern Renaissance portal were left. The beautiful interior, in which concerts are held because of the excellent acoustics, is closed off by a raised choir. Here you can still see remnants of Romanesque ceiling paintings. To the right of the choir you can look through a grille into the Ludmilla Chapel, with the tomb of the saint, the grandmother of Prince Wenceslas. The tombs of two Bohemian nobles are placed in front of the choir. The Baroque statue in front of the crypt—a corpse with snakes in its intestines—is a realistically portrayed allegory of the transitory nature of life.

The Baroque **Chapel of St. John Nepomuk** is incorporated into the outer facade of the basilica. Its portal is decorated with a statue of the saint by F.M. Brokoff.

Adjoining the basilica on the left is the former Benedictine **Monastery of St. George** (founded 973), rebuilt several times, which today houses the **Old Bohemian Art Collection** (part of the National Gallery). On exhibition here are works from the 14th to the 18th centuries, among them pictures by artists who took part in the building of many of Prague's churches.

Along the north side of St. Vitus' Cathedral runs the Vikárská ul., in

Skilful wrought iron work—part of St. Vitus' Cathedral.

which there is nowadays a tourist information office. Nearby is the **Milhulka Powder Tower**, which in recent years has been opened to the public. In the late 15th century, while parts of the northern fortifications were being built, it served as a gunpowder store. Today it contains a small museum, which records the tower's earlier use as a metal casting foundry and possibly an alchemist's laboratory. Individual storeys portray aspects of crafts in the 16th and 17th centuries.

Goldmakers' Alley

Another part of the fortifications can be seen behind St. George's monastery. Go past the basilica and turn left up the **Goldmakers' Alley** (*Zlatá ulic ká*), considered to be one of the most popular attractions of the castle. In the part of the fortifications between the central **White Tower** and the

outermost **Daliborka Tower**, tiny houses crouch under the walkway on the wall, making a romantic backdrop. Legend has it that this is where the famous alchemists employed by Rudolf II tried to discover the secret of eternal life and how to make gold. It is a fact, however, that Franz Kafka lived and worked on his novels at times in house no. 22.

The tour of the castle ends at the **Black Tower**, where the Jiřská ul. reaches the eastern gate. Just before the gate, on the right-hand side, you come to the **Palais Lobkowic**, which in recent years has been used for many exhibitions based on themes from the country's history. On the other side of the eastern gate the **Old Castle Steps** and the street Na Opyši lead down to the Malá Strana and the underground railway stop, Malostranská.

The Belvedere Palace

Lying outside the castle complex, the **Belvedere Palace** (*Kráovsky letohrádek*) is worth visiting. Leave the Second Courtyard and go north, cross a bridge (*Prašy most*) going towards the former **Riding School** (*Jídárna*), nowadays an exhibition hall. From its terrace you can see an imposing view of the St. Vitus' Cathedral. Leaving the Riding School, follow the **St. Mary's Wall** (*Mariánské hradby*), which borders the Royal Gardens (unfortunately closed to the public) round to your right. It is not far now to the splendid palace, which art historians consider to be the only example of a purely Italian Renaissance building north of the Alps. The Emperor Ferdinand I had it built in the mid-16th century for his wife Anna. Especially remarkable is the "singing" Renaissance fountain in the garden, which was cast with such skill that the falling water, striking the bronze basin decorated with hunting scenes, makes it ring.

Left, reflected in the water of the fountain—Goldmaker's Alley. Right, wild vines on the Old Castle Steps.

THE GOLEM—A MYTH IS BORN

Every city with a long history has its stories, in which real events are shrouded in mysterious shadow. Prague has such legends even about its founder, the Princess Libussa. There is also the legend of the headless Templar knight who rides through Prague, which dates from the persecution of the Templar order. And there are the water spirits who guard the many dams of the Vltava. There is also Doctor Faustus, who tried to make gold, using a pact made with the devil, no less. Finally the Prince of Dark-

have been reached, but then new doubts, new uncertainties arose. The new knowledge blundered on into the uncertainties—searching, believing, doubting, mistaking, often falling into despair.

And yet the Golem, a fantastical-mechanical monster, an animated piece of mere matter, could really only have come into existence in the mysterious ghetto of the time of Rudolf II. It was this Roman Catholic emperor and Bohemian king in the midst of a country of heretics who created the image

ness dragged him off to his kingdom forever, drawing him through the ceiling of the Faustus house. This legend is probably based on the figure of the alchemist Mladota, who lived in Prague.

However, none of these tales ever obscured the real, historical Prague. Not until the late 16th century did an image of a fantasy Prague begin to form. These were strange times. The glory of the Renaissance was giving way to the twilight of the Baroque. The Renaissance had freed humanity of the superstition of medieval times, but was not yet able to force its discovery of reason on the world. Certainty seemed to

of this fantasy Prague, a city of alchemists and artists, of astrologers and scholars, who tried to lift the veil of divine secrets. A world of dreamers, of deluded seekers for truth and of lost pilgrims in search of eternal verities. This was the atmosphere of the city in which the miracle-working Rabbi Löw and his Golem could meet up with the most famous magician of his time, Doctor Faustus.

A fantasy Prague, whose atmosphere of wonder magic and rich legend cannot be equaled by any other epoch. Not even by our own times, when the discoveries of the secrets of nuclear physics have already outstripped all the theoretical mysteries of the sages and magicians of Rudolf's era.

Here then is the legend of the Golem, the creature of mud and clay made by the cabalist, astronomer and magician Rabbi Löw, who breathed life into the Golem with a magic word, the "Shem", "in order to send it out to protect his community, to discover crimes and to prevent them", as the *Sippurim*, a 19th-century collection of Jewish legends, tells. One evening, before the Sabbath rest, Rabbi Löw forgot to remove the sign of life from the mouth of the Golem, and

being. Not until the commentary of Eliezer of Worms, dating from the 13th century, does the word "Golem" in the sense of an artificial creation appear (along with exact instructions for making such a creature). In the medieval stories, the Golem is portrayed as a perfect servant, its only fault being that it interprets its master's instructions too literally. By the 16th century, it was seen as a figure that protected the Jews from persecution, but it had also acquired a sinister aspect. Not until the middle of the 19th century

the latter began a rampage of destruction in Löw's house. The spark of life was removed and the creature turned back into mud and clay, to lie forever under the roof of the Old New Synagogue.

The origins of the legend itself are based on the first part of the Kabbala, the mystical teachings and writings which are already mentioned in the Talmud, but which contain no word of an artificially created human

do we find any connection in writing between these tales of the "creative" Rabbi and the figure of Rabbi Löw the alchemist. According to this version, Rabbi Löw, clothed in white, went one dark night to the banks of the Vltava and there, with the help of his son-in-law, he created the Golem, while continually chanting spells, from the four natural elements—earth, fire, air and water.

This story forms the basis of the German novel *Der Golem* by Gustav Meyrinck. In the 1920 it was made into a film, which became a classic of German silent cinema, and the Golem became the model for many other cinematic man-made monsters.

Left, tombstones in the old Jewish cemetery. Above, the Star of David, symbol of the Jewish faith, located above the Jewish Town Hall in Maiselova.

THE STRAHOV MONASTERY

It lies outside the castle fortifications and outside the whole castle complex, on the age-old trade route from Nuremberg to Krakow, on the slopes of the Petřín Hill, the crown of the gently sloping valley. It is the **Strahov monastery**, the oldest Premonstratensian monastery in all Bohemia, now lying on the Strahovské nár. The two towers of the Strahov, along with the green of the Petřín hill and its miniature "Eiffel" tower, and the long line of the roof of the Palais Czernin, all make up the unmistakable silhouette of the left bank of the Vltava.

The first monastery of the white monks of the Premonstratensian order was founded in 1140 by King Vladislav II and remained in existence—apart from a few historical interruptions, for instance during the Hussite wars—until 1952. After the dissolution of all religious orders in the CSSR, the Strahov monastery was declared a museum of national literature and opened to the public in this guise on May 8, 1953. This rapid changeover was possible because of the resources of the monastery library, gathered over the centuries and among the most select collections in the country. Strahov possesses not only the oldest and most extensive, but above all the most valuable library in the country. The core of the collection was established at the time of the foundation 800 years ago. Gradually added over the years were examples of almost the complete literature of western Christianity up to the end of the 18th century. Today the emphasis is on national literature of the 19th and 20th centuries.

If you go into the harmonious enclosure of the monastery, the first thing you will see is the church of St. Rochus, built from 1603-1612 during the rule of Emperor Rudolf II, and nowadays used as a gallery. On the facade of the "New" Library (built from 1782-1784) is a medallion with the portrait of Emperor Joseph II, the ruler whose support of the Enlightenment led to the dissolution of the majority of monasteries in his domains. His memory is honored here because he allowed Strahov to remain, and the monks of Strahov to buy the complete inventory for the building of a new library from another famous monastery in Moravia, the Bruck monastery near Znojmo. These brown and gold gleaming shelves equipped the new hall, which was then designated the "Philosophical Hall". The older hall was renamed the "Theological Hall".

Theological and Philosophical Halls

The **Theological Hall** was built by Giovanni Domenico Orsi in a rich Baroque style at a cost of 2254 guilders. It was painted with splendid ceiling frescos by Siardus Nosecky, a member of the monastery, from 1723 to 1727. The theme is true wisdom, rooted deeply in the knowledge of God. The brightly colored scenes in their sturdy stucco frames radiate warmth, freshness and cheerfulness.

The ceiling fresco in the **Philosophical Hall** is less easy to read. It is in concept and technique a monumental finale to Baroque ceiling painting in Europe. The fresco in Strahov shows the development of humanity through wisdom—a theme that borders on the ideas of the Enlightenment. On the two narrow sides you can see Moses with the tablets of law and, opposite, Paul preaching at the pagan altar. The long lines of figures on the long sides introduce the great personalities of history who have made tremendous progress possible through their achievements.

Preceding pages: the Theological Hall in the library of the Strahov monastery.

The Strahov Library

In 1950, the library contained 130,000 books. By now this number has grown to around 900,000, as the Strahov has taken in the contents of a number of other monasterial libraries, particularly in central and northern Bohemia. One of the most famous of all illuminated manuscripts is the Strahov Gospels, the oldest manuscript in the library, dating from the 9th to 10th century during the reign of Charles IV. A copy can be seen in the Strahov. Also among the most valuable treasures are rarities such as the New Testament, printed in Pilsen in 1476 and one of the first printed works in Czech, or the beautifully illustrated story of the journey of Frederick von Dohna to Rome, dating from the 17th century.

A great deal of this can be seen in the photographs and transparencies that form part of a special cultural and historical tour set up in the cloisters. A separate room is devoted to the great reformer, Jan Hus.

The upper storey contains a great number of documents relating to the writers who influenced the re-awakening of Czech national consciousness in the 19th century.

Church and Monastery Gardens

If the church happens to be open, it is best not to miss the great opportunity to visit and admire this originally Romanesque building, which was vastly altered and richly redecorated in the 17th and 18th centuries.

Also part of the monastery grounds are the large monastery gardens. These fill the valley between the Petřín and the castle hills right up to the edge of the Malá Strana. Once white-robed monks walked here, nowadays it is a favorite refuge for courting couples.

View over the rooftops of the Malá Strana.

THE VLTAVA

The Vltava is 270 miles (435 km) long and rises from two headstreams—the Tepla Vltava, rising on the mountain of Cerna hora, and the Studena Vltava. It flows first south-east, then north across Bohemia. Once a wild river whose floods threatened the city, the Vltava today flows peacefully and calmly. Dams—among them three large hydrodams with lakes that are also used for recreation—raise the water level and slow down the current, and the Vltava gives an impression of might and majesty. When it

Boats for river excursions leave from the Palacky Bridge. They travel in both directions, downstream to the zoo and the Prague suburb of Roztoky, upstream to the artificial lake of Slapy. Rowing boats can be hired and are very popular—it is impossible to imagine the Vltava in summer without them. The winter fairs on the frozen river are unfortunately a thing of the past, as the damming of the river has warmed the water and the river no longer freezes. The oldest open-air swimming pool on the Vltava dates from 1840 and

reaches the town of Mělník, it joins the Elbe, called Labe in Czech, (the latter only becomes navigable from this point on), and by this route connects Prague with the West German seaport of Hamburg. Thus Vltava forms an important link for river traffic between Czechoslovakia and the North Sea. The **Prague river harbor** lies in the bend of the Vltava in the city district of Holesovice. Nothing remains of the earlier harbors which once lay on a tributary (now filled in) of the Vltava except the pub "Hamburk" in the square of Karlínské nám.

Apart from freight and shipping, the river is mostly used for recreation and enjoyment.

is situated on the Malá Strana bank next to the S. Cech Bridge.

From the hills of the left bank you can get a beautiful view of the arrangement of the bridges, which lie one behind the other. For 500 years the old Charles Bridge was the only link between the two river banks. Not until the mid-19th century Industrial Revolution were more bridges built. At about this time, a start was made on building the embankments. Only the Malá Strana bank in the vicinity of the Charles Bridge has been altered in this way. Kampa Island and the mouth of the Certovka tributary have remained in their original state.

The Vltava has been honored in various different ways. The allegorical representation of the Vltava is popular—it's a statue which decorates a fountain on the exterior of the garden wall of the Palais Clam-Gallas. Its popular name is **Terezka**—rumor has it that a wealthy citizen of Prague left his whole fortune to the statue. In the late 18th century, at the time of the birth of the Czech nationalist movement, the Vltava was a never-failing source of inspiration. The Vyšehrad castle on its rock above the Vltava was famed in

bank of the Vltava, not far from the Vyšehrad rock, where it forms another dominant feature of the city skyline. The theater was opened in 1881 with a performance of Smetana's *Libussa*, in which the solemn prophesy of the mythological princess is heard. According to another legend, mostly omitted from the artistic versions, Libussa had a very prosaic relationship with the river. It is claimed that, when she had enough of her lovers, she had them thrown from the Vyšehrad rock into the Vltava. The national

many legends, and in the Romantic period it fascinated many Czech artists searching for their national identity. The myths and the river that was linked with them found their expression in countless songs, works of representational art and literature. Later, this tradition was continued by a whole generation of artists who took part in the construction of the National Theater, that symbol of the completion of national rebirth. The building was, appropriately, sited on the

Preceding pages: Charles Bridge. Left, once upon a time—winter amusements on the frozen Vltava. Above, spring flooding.

composer Smetana took up the Vyšehrad myth once again when he composed the symphonic poem *Ma Vlast* (My Homeland). The second movement deals with the Vltava and is perhaps the most famous artistic representation of the river. Even today, the river still influences people's imagination, though in a quite different way. Children in particular are familiar with the Vltava water spirits which appear in fairy tales. The Czech water spirits, little men with green coats and pipes, have lived in the water since time immemorial. The water spirits know every stone and every fish, are very wise and are always ready to give good advice.

MALA STRANA

Malá Strana lies at the feet of the castle of Prague. It is a totally individual quarter, a picturesque island, separated from the rest of the noisy big city by broad parks and the wide, steady flow of the Vltava. Looking down from the hills, the impression is of a landslide of roofs which started to roll between the Hradčany and Petřín Hills and came to a stop on the river bank.

In 1257, Malá Strana was made a city, and is thus the second oldest among the four historic cities that make up Prague. The Malá Strana experienced its first boom during the rule of Charles IV. During this time it was extended considerably and received new fortifications. However, not until catastrophic damage was done by the great fire of 1541 was there any sign of a major rebuilding program. The rebuilding after the fire shaped the individual characteristics of this quarter and we can still see them today.

The Malá Strana truly blossomed after the victory of the Catholic League in the Battle of the White Mountain (1620), when many wealthy families loyal to the House of Habsburg settled here. True, most of the palaces were deserted once the political administration of Bohemia had moved to Vienna, but the palaces have been spared major alteration to this day. Even the citizens' houses, which often have older foundations, have kept their mainly Baroque facades with their characteristic house signs. For this particular reason, the Malá Strana can be described as an architectural jewel, indeed as a complete work of art representing the Baroque style of Central Europe.

The different creative styles of the citizens' houses, the small, quiet

View of St. Nicholas and the roofs of the Malá Strana.

squares, the palaces with their gardens designed to fit into the hill slopes all came together to form an original style—"Prague Baroque".

Of course, many things worth seeing remain hidden behind the facades. However, once off the main streets, you can enjoy the special atmosphere of the place. This is not least due to the fact that by no means everything revolves around the tourist, and the Malá Strana has its own everyday life to live.

Around Malostranské náměstí

The center of the Malá Strana always was and still is the Malostranské náměstí, a square which is actually divided into two squares by St. Nicholas' Church (*Chrám sv. Mikuláše*) and the neighboring former Jesuit college.

The conspicuous dome of the St. Nicholas Church and its slender tower can be seen from many different view-points in an ever-changing perspective. This unequal couple have become the symbol of the whole Malá Strana. The church itself is a masterpiece of Baroque architecture and one of the most beautiful examples of its kind. In the early 18th century the famous architect Christoph Dientzenhofer built the nave and side chapels on the site of a Gothic church. The choir and the dome were added later by his son Kilian Ignaz. The building was completed in the mid 18th century by the addition of the tower, the work of Lurago.

Particularly outstanding among the special features of the interior is the monumental ceiling fresco by J.L. Kraker in the nave. It is one of the biggest in Europe and portrays scenes from the life of St. Nicholas. Another valuable fresco, by F.X. Palko, decorates the dome. The dome is 247 feet (75 meters) high—tall enough to accommodate the tower on the Petřín hill

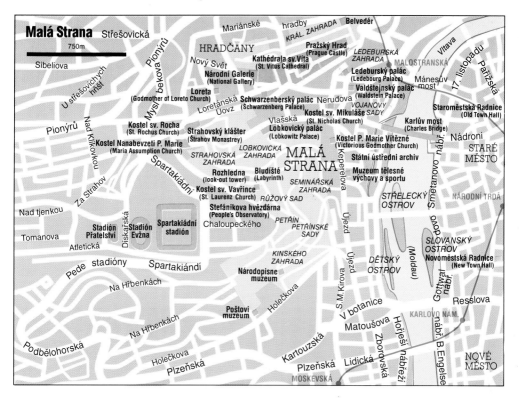

inside it. The sculptures in the choir and the gilded statue of St. Nicholas, patron saint of merchants and sailors, are the work of Ignaz F. Platzer the Elder.

Opposite the church is the **Palais Lichtenstein** with its broad Neo-classicist facade. From 1620 to 1627 it belonged to Karl von Lichtenstein, the so-called "Bloody Governor" who was mainly responsible for the execution of the leaders of the 1618 rebellion.

From St. Nicholas' Church you will notice the **Golden Lion** house (*U zlatého lva*). It is one of the few purely Renaissance houses in the Malá Strana and also contains the wine bar *U mecenáše*. These small wine bars, whose charm lie mainly in their ancient walls, are typical of the Malá Strana. Guests were served here as early as 1600. Nowadays, however, it's not always easy to find a seat. For beer drinkers, the pub **U Glaubicu** has always been a household word. The corner house of the same name, a little further on under the arcaded passage, is being renovated. We will have to wait and see if the tradition of excellent beer continues after the renovation.

Cross the street **Karmelitská** and follow the arcaded passage past a back courtyard rich in atmosphere. It lies hidden behind the arch of a gateway. All over the quarter, little surprises such as this await the observant visitor. However, visitors should cultivate a sense of tact, because even the proverbial hospitality of the Czechs, which is under considerable strain during the tourist season, has its limits.

The lower side of the square is bordered on the right by the **Palais Kaiserstein**. A memorial plaque inside the house proclaims that the world-famous opera singer Emmy Destinn once lived here. On the left side is the former town hall of the Malá Strana (it is inscribed "Malostranská beseda"). Today it's the place to go for anyone who wants to hear jazz in Prague. The well-estab-

lished **Café Malostranská kavárna** protrudes into the square and, in the summer months, provides one of the few opportunities for eating and drinking out of doors.

Leave the lively Mostecká ul .and on your right, you will find that the streets have become quieter. Here you enter one of the quiet, dreamy corners of the Malá Strana. There is a passage next to the cinema *"U hradeb"*, or you can turn off into the Lázeňská ul. House no. 6, *"Vláznich"* (At the Baths) was a first class hotel in the 19th century. It was Tsar Peter the Great, for example, stayed here. A memorial plaque proclaims that the French poet René Chateaubriand also stayed here. Another memorial plaque, on the house *"U zlatého jednorožce"* (The Golden Unicorn), marks the place where Ludwig van Beethoven stayed.

You can find a beautiful Baroque interior in the church **St. Mary beneath the Chain** (Kostel Panny Marie pod ř etezem), the oldest church in the Malá Strana. Remains of the walls of its predecessor, a 12th century Romanesque basilica, can still be seen in the right-hand wall of the forecourt.

An excursion to your right will take you to the long sprawl of the **Maltese Square** *(Maltézské nám.)*, with its sculpture representing John the Baptist, which is the work of F.M. Brokoff. Two palaces in this square are worth your attention: the **Palais Turba** (now the Japanese Embassy) with its Rococo facade, and the Early Baroque (later rebuilt) **Palais Nostitz** (Dutch Embassy, Ministry of Culture) which closes off the square to the south.

Located next to St. Mary's church another little square is joined on—the Velkopřevorské nám. On the one side is the **Palais Buquoy**, home of the French Embassy, and opposite is the former **palace of the Grand Prior of the Knights of Malta**, one of the most beautiful in the Malá Strana. Today it

Left, trams in the Letenská.

contains the **Museum of Musical Instruments**, and its vast collection is worth seeing, as it is of interest even to non-musicians. In the summer, numerous concerts are held in the adjoining "Maltese Gardens".

A little bridge connects the square with the island of Kampa. It is separated from the Malá Strana by a tributary of the Vltava, the *"Certovka"*. A little further upstream you can discover an old millwheel. The park has been formed by linking up the gardens of former palaces and offers a beautiful view of the Old Town. Between the Charles Bridge and the mouth of the *"Certovka"* there is a small group of houses, lying directly by the water, which are popularly known by the proud name of "the Venice of Prague".

Go up a double set of steps and you will come up onto the Charles Bridge *(Karluv most)* which, together with the silhouette of the Hradčany castle, has

become the symbol of Prague. A wooden bridge, linking the two banks of the Vltava at approximately the same place, is mentioned as early as the 10th century. In 1165 it was probably replaced by the stone Judith Bridge, the second oldest stone bridge in central Europe. After the destruction of the Judith Bridge in a flood, Emperor Charles IV had a new one built by his cathedral architect Peter Parler. It received the emperor's name. This bridge was also damaged by floodwater several times, but never collapsed. According to a persistent legend, eggs were mixed with the mortar to give it durability. Perhaps, also, it is no coincidence that the foundation stone was laid on July 9, the day of Saturn's conjunction with the sun. In any case, the astrologers, who in those times were often consulted when important decisions were to be made, considered it to be a most auspicious moment. These

and other theories have often been cited to try and discover the secret of the bridge. Whatever it may have been, this 600-year-old construction is deserving of admiration, especially as it even stood up to the car and tram traffic of the 20th century, until it was declared a pedestrian precinct and restored to its well-earned rest.

Statues on the Bridge

The mainly Baroque statues (nowadays partially replaced by copies) were created in the late 17th century after the model of the Ponte Sant'Angelo in Rome. In contrast to the Gothic architecture, these give the bridge its characteristic appearance. The famous artist Johann Brokoff, his sons Ferdinand Maximilian, Michael Josef, and Matthias B. Braun were among those who worked on the statues. The last named was responsible for what is perhaps

The route from the Charles Bridge to Malá Strana Square crosses the Mostecká.

artistically the most valuable sculpture: the **statue group of St. Luitgard** (1710). It shows Christ appearing to the blind saint and allowing her to kiss his wounds.

The oldest statue on the bridge is that of **St. John Nepomuk** (1683), designed by M. Rauchmuller and Johann Brokoff. The reliefs around the base portray scenes from the life of the cleric, canonized in 1729. One relief is based on the legend of John Nepomuk's death: the wife of Wenceslas IV (one of the sons of Charles IV) had made her confession to Nepomuk, whereupon the king put pressure on him to disclose the details. John Nepomuk steadfastly refused to break the seal of the confessional, and in 1393 Wenceslas IV had him drowned not far from this spot. In actual fact his belated canonization probably had more to do with the intention to oust the memory of Jan Hus with the cult of a new saint.

Another popular figure is the **statue of Bruncvik** on one of the bridge supports on the banks of the island of Kampa. The sword of this legendary knight, who is associated with the Roland epic, is—so it's said—walled up in the bridge and can be retrieved in the country's darkest hour.

At the end of the bridge are the **Malá Strana Bridge Towers**, with an archway in the middle. Among the coats of arms that decorate it is that of the Old Town. The bridge, including the Malá Strana towers, actually belonged to the Old Town. That is why the former customs house can be found on this side of the bridge, on the left in front of the gate. The smaller tower is a remnant of the Judith Bridge, and only its Renaissance gables and the wall ornaments were later added. The higher tower, dating from the 15th century, was designed to complement the Old Town tower. The top is open to the public. If you look down towards the Old Town, you will see why Prague is also known as the "city of a hundred spires".

Downstream from the Charles Bridge the palace gardens of the Malá Strana beckon. Some of them are open to the public in the summer months. Without leaving the bridge you can see the Renaissance house *U tří pš rosu* (The Three Ostriches). The remains of its sgraffito decoration show that it once belong to a supplier of feathers to the royal house. This house points the way to the street U lužického semináře, where the former **monastery garden** *(Vojanovy sady)* is situated. In this park with its two Baroque chapels modern sculptures are often exhibited. You certainly should not miss the chance to visit the **Waldstein Garden**, which can be reached from the Letenská ul.. We shall refer to it again later in connection with the Palais Waldstein.

If you not only know about Czech beer (the dark beer, in this case), but consider it to be an essential, you should visit the traditional inn **U sv. Tomáše** (St. Thomas') in the same street. The beer garden of this former monastic brewery (founded 1358) is unfortunately no longer in existence, and neither is the home brew, but Bráník beer tastes no less excellent down in the cellar vaults. A church is of course also part of this former Augustinian monastery. It was founded in the 13th century. Its present Baroque form is the work of Kilian Dientzenhofer, and inside it is richly decorated with the works of Bohemian Baroque artists.

Malá Strana Gardens—Palais Waldstein

When making a tour of the Malá Strana gardens, you should also take a quick look at the Underground station Malostranská, which contains a copy of M.B. Braun's *Hope*. Some sculptures from his workshop can be seen in

Kayak training in a nearby stream.

the courtyard garden. Your route now takes you on through the Valdštejnská ul. past palaces whose gardens lie on the slope beneath the castle, an ideal place for artists. Three of these terraced gardens, built for the nobility after Italian models, can be visited. The entrance is next to the **Palais Kolovrat** (no. 10).

The Valdštejnská ul. and Waldstein Square *(Valdštejnské nám.)*, into which it leads, border the broad complex of the **Palais Waldstein** on two sides. This first Baroque palace in Prague was built between 1624-30 for the famous (and infamous) general Albrecht von Wallenstein and was a worthy memorial to his ambition. This hero of Schiller's play *Wallenstein* had made his way to the top by skilful strategy and leadership on the one hand and by intrigues and treachery on the other. During the Thirty Years' War he enlisted under the Habsburg Ferdinand II. He won many important victories for Ferdinand II. These brought the imperial generalissimo not only power but also, along with a ducal title, wealth, which as a court favorite he was particularly well placed to increase. This process was helped not least by his participation in a grandiose coin swindle, so that in the end he was able to raise his own private army. His rapid rise came to an equally rapid end when, in secret deals with the enemy, he initiated tactical manoeuvres which would, in the end, have led him to the Bohemian crown. However, the emperor saw through him and had him murdered in Eger in 1634.

The grandiose residence matches Wallenstein's political ambitions. It was intended to rival the Prague castle. He gained the site for the building by buying up and dispossessing the inhabitants of more than 20 houses. Even the city gate had to go, in order to give the architects (all Italians) enough space to provide their patron with a palace featuring all possible luxuries available at the time.

However, the rather restrained outer facade that faces the square does not give anywhere near the same impression as a visit to the palace gardens mentioned above. The greatest pride of the householder—apart from an artificial grotto, an aviary and a pond—was the triple arched **loggia** *(sala terrena)*, richly decorated with frescos. Today it serves as the podium for open-air concerts. The bronze statues of mythological gods and goddesses, scattered about the garden, are the work of Adriaen de Vries, court sculptor to the Emperor Rudolf II. They are, however, copies—the originals were taken to Sweden as spoils of war (1648) and now it is located in the park of the Drottingholm palace near Stockholm. Another work by this Dutch sculptor is the figure of Hercules fighting the dragon in the middle of the small pond. The fountain with the sculpture of Venus and Cupid is also remarkable.

At Eastertime, painted eggs are on offer everywhere in Prague.

Opposite the loggia is the former **Riding School**, where art exhibitions are held nowadays.

The Tomášska ul. leads from Waldstein Square back to Malá Strana Square. Go past the house "The Golden Bretzel" (no. 12) and you will come to the Baroque house "The Golden Stag". This house bears one of the most beautiful house signs in the whole of Prague. It shows St. Hubert with a stag. The sculpture is the work of F. M. Brokoff.

Before house numbers were introduced, during the reign of Maria Theresia in the 18th century, these house signs were used to identify the houses. They were based on the profession or craft of the house owner, his status or the immediate environment of the house. Animal and other symbolic signs, both of a secular and a religious nature, were popular. If the owner changed, the house retained its original sign. Sometimes the new owner even took over the name of the house.

We recommend that you go back to Waldstein Square and from there, make a short excursion into the Sněmovní ul. This street and the adjoining cul-de-sac, *U zlaté studně* (The Golden Fountain) form a picturesque corner. Hidden away at the end of the little alley is a garden pub with the same name as the street. Also noteworthy is the Renaissance house **The Golden Swan** (*Sněmovní ul.* no. 10), which hides a beautiful inner courtyard. Go back in the direction of the Thunovská ul., which leads into the **Castle Steps**. These so-called New Castle Steps are not to be confused with the Old, which lead to the other end of the castle. The New are much older than the Old, but that's the way of things in Prague.

Neruda Alley

Parallel to the Castle Steps (Zámecké schody) lies the Neruda Alley *(Nerudova ul.)*, named after the famed Czech poet, author and journalist Jan Neruda (1834-1891) who lived in the upper part of the street, in house no. 47, **The Two Suns**. His work is inspired by the everyday life of the Malá Strana. Incidentally, his name was adopted by the Chilean poet Ricardo Eliecer Neftalí Reyes y Besoalto, now the Nobel Prize winner Pablo Neruda.

Many of the middle class houses in this street were originally built in a Renaissance style and later given Baroque additions. They often bear house signs which don't match the names of the houses. For instance, house no. 6, **The Red Eagle**, has a sign showing two angels. In the case of house no. 12, **The Three Violins**, it is known that several generations of violin makers lived here. More signs can be seen on the houses **The Golden Chalice** (no. 16), **St. John Nepomuk** (no. 18), **The Donkey and the Cradle** (no. 25) and others. There is a small pharmacological museum in the

Left, all sorts of different views of Prague are for sale at the Charles Bridge. Right, typical steps in the Malá Strana.

former pharmacy **The Golden Lion.**

As is so often the case in Prague, two embassies have settled into the two Baroque palaces in this street. On the left is the **Palais Morzin** (Rumanian Embassy). Its unusual facade ornament—the heraldic Moors which support the balcony, the allegorical figures of Day and Night and the sculptures representing the four corners of the world—are the work of F.M. Brokoff. Somewhat higher up is the **Palais Thun-Hohenstein** (Italian Embassy), which is decorated with two eagles with outspread wings, and is the work of M.B. Braun. The two statues of Roman deities represent Jupiter and Juno.

The palace is connected to the neighboring **church of St. Kajetán** by two passages, beneath which the stairs lead up to the New Palace Steps. This link continues, diagonally opposite, to the Tržiště street.

From the end of the Neruda Alley you can get a splendid view of the **Palais Schwarzenberg**, mentioned above in connection with Hradčany Square. To get there, you either go right across the Castle Ramp or left by means of the **Town Hall Steps**. If you go straight on, you leave the Malá Strana in the direction of the **Strahov monastery.**

At the back of the last houses of the Neruda Alley a maze of courtyards lies hidden. They fall in a series of terraces into the valley between the two hills. At the bottom are a few alleys that have almost a village character. If you go back a little, you will reach the Rococo **Palais Bretfeld** (no. 33), with a relief of St. Nicholas on the portal. In earlier years famous balls took place in this building—Giacomo Casanova is claimed to have attended them.

From here the steps Jánsky vršek lead down and then turn right into the Sporkova ul., which leads us along the slope mentioned above. It then curves and leads into the Vlašská ul., directly opposite the **Palais Lobkowic**. This

magnificent Baroque palace now contains yet another embassy—the West German Embassy. The palace garden is partially open to the public and is worth visiting because of the view.

If you want to visit the parks of the Petřín hill or the Strahov monastery, you have to follow the the Vlašská ul. to your right. However, you can also go on in the opposite direction and use the cable car, which runs between the Ujezd street and the peak of the Petřín Hill. On the way lies the **Palais Schönborn** which now houses the US Embassy. Its splendid garden, which can be seen from the Castle Ramp, is not open to the public. However, you can visit the particularly lovely Baroque terraced garden of the **Palais Vrtba** (*Karmelitská ul*. no. 25). It is a small garden, but full of atmosphere, and has a *sala terrena* and sculptures by M. B. Braun.

Further along the Karmelitská you come to the **church of St. Mary of the Victories** (*Kostel Panny Marie Vitezné*), the first Baroque church to be built in Prague. It was built as a monument to the Counter-Reformation brought to Prague by the Habsburgs. The furnishings, still a stylistic unit, date from the 17th century, the saints' pictures by the altar are by P. Brandl. In this church the **Infant Jesus of Prague** is kept. It is revered and believed to work miracles, and has achieved worldwide fame. It is a wax figure of Spanish origin and is always clothed in one or other of its 39 costly robes.

The Petřín Hill

It's not far from here to the **Petřín Cable Car** mentioned above. It also stops at the restaurant "Nebozizek", which has a magnificent view. It is known as a cable car, but the cars don't hang on cables, they run on rails. This curious construction is based on the original motor system. The old cars had a water tank, which was always filled at the top and emptied at the bottom. In this way the cars going up were powered solely by the weight of the cars going down. This system opened in 1891. In the 60s the old system was demolished and replaced by a modern construction.

The park on the Petřín (St. Laurence's Hill) was formed by linking up the gardens which had gradually replaced the former vineyards. On the level of the upper cable car station, a path offering a marvelous view leads all the way through the park to the Strahov monastery. Apart from the delightful view, the park also has other sights to offer.

Starting in the most southerly corner, visit the **Villa Kinsky** which houses the Ethnographic Museum. On the way, in a northerly direction, lies a little wooden church, a wonderful example of folk art of the 18th century. It comes from the Carpathian Ukraine and was

Palais Schönborn: nowadays seat of the US Embassy.

rebuilt on this spot in 1929. The church was a gift of the inhabitants of a small village in that region, which became part of Czechoslovakia after World War I and part of the Soviet Union after World War II.

The **"Wall of Hunger"**, which you will discover further on, leads down the slope and is part of the fortifications which Charles IV had built. According to rumor, this project was undertaken to provide work for the starving and impoverished. Near the wall lies the **People's Observatory**, a popular meeting place of amateur astronomers from near and far.

On top of the hill is the **Viewing Tower**, a copy of the Eiffel Tower, which is 197 feet (60 meters) high. It was built for the Prague Jubilee Exhibition in 1891. Not far from the tower is **St. Laurence's Church** and a **labyrinth of mirrors**. The latter brings the tour to a pleasant end, particularly for children.

The claim is often made that Bohemian cuisine is anything but healthy or easily digestible for Western stomachs. Once you have accepted the fact that vegetables and salads are in short supply, that in the plainer pubs and restaurants the relationship of meat to dumplings is about one to four, that paprika and pickled Prague gherkins are the standard accompaniment to any dish, and that boiled Prague ham is on offer almost everywhere in a hundred different varieties, you can safely go and explore the Prague *pečeně)* with Bohemian dumplings *(knedlíky)* and cabbage *(kyselé zelí)*? For quick service, value for money and a good meal go to **U Bonaparta**, Prague 1, Nerudova 29. With it, drink a beer (specific gravity 12 degrees) from Smichov. Also in the Bohemian tradition, only kosher, is the food served in the **Jewish Restaurant** (3rd skupina, or price bracket) in Prague 1, Maiselova 18.

If you want a more formal meal, with an aperitif and several courses, then you would

restaurants. Only one more hurdle remains to be overcome before enjoying a hearty Bohemian meal—the waiter. It can happen that all the tables are "reserved", although no one is sitting anywhere, and there's no sign of anyone coming, either.

All the large hotels in Prague have several restaurants and snack bars. Each hotel in the top price bracket also has both a Bohemian and a French restaurant.

Bohemian Cuisine

However, where should you go to eat genuine Bohemian roast pork *(vepřová* do well to go to the **Pelican**, Prague 1, Na příkopě 7. However, in restaurants of the 2nd skupina such as this one, it is necessary to reserve a table. The decor is not over-lavish and the guests are not overloaded with enormous white napkins, but can relax and enjoy good food and Moravian wines. If your foreign language skills extend to German, but not Czech, the waiters will translate the menu to fluent German. This service cannot be match by the **Staropraszká Rychta**, in the cellar of Wenceslas Square 7, specializes in old Czech dishes. The guests in the adjoining hotels Zlatá Husa and Ambassador fear the celebrations that follow the weekly slaughtering day, mostly Thursday. Right next door and also in the cellar is the **Halali-Grill** (Wenceslas Square 5). It is in the 1st skupina and offers game specialities and a gypsy band. Also recommended for game specialities is the **Myslivna** (2nd skupina) in Prague 3, Jagellonská 21. If you would like fish for a change, we recommend the **Baltic Grill** in the passage at Wenceslas Square 43.

The people of Prague like their food flambéed, and among the masters of this

eat an excellent meal in the **Lobkovická vinárna**, Prag 1, Malá Strana, Vlašska 17.

Fast Food

The recently opened **arbet** now caters for fast food fans. Further details are unnecessary, all you have to do is to follow the trail of red and gold French Fries tubs that litter the Na přikopě. And if you like good, solid food, you can buy a fried sausage from the stalls in Wenceslas Square, which

skill are the waiters in the former Ursuline convent in the Národní tř., the **Klášterní vinárna**. Here the delicious *palačinky*, thin pancakes filled with ice cream and fruit, are turned once more in blazing alcohol. It lies right next to the new theater and the National Theater, and as such is a good address for a late supper after a visit to the opera. Surrounded by American, German and Italian diplomats, you can—how could it be otherwise among such a select clientele—

Left, an old Prague beer cellar in pre-1900 times. Above, breakfast tables in the Hotel Europa.

are open until late at night.

If you've had enough of dumplings and pork, the **Trattoria Viola** (2nd skupina), Národní tř. 7, offers a chance of escape. Here you will find an interesting combination of Italian cuisine and limited Bohemian ingredients. One further point in its favor: it serves Chianti wine. Indian cuisine can be enjoyed in the extremely formal top-class **Indická restaurace**, Prague 1, Nové mesto, Stěpánská 63. Also particularly popular is the **Cínská restaurace**, Prague 1, Nové mesto, Vodičkova 19. However, it is not unusual to have to book three or four days in advance here.

STARÉ MESTO—THE OLD TOWN

The Old Town (Staré Město) of Prague is spread along the right bank of the Vltava and around the Old Town Square. The main streets—Národní třída, Na přikopě and Revoluční—mostly follow the course of the city fortifications, which no longer exist. As the name *"Na přikopě"* which means By the moat indicates that they were built on the site of the old moat which separated the Old Town from the New. Together these two districts form the actual city center of Prague.

However, the Old Town has kept its individual character. The pattern of streets and squares has in the main stayed unaltered since the Middle Ages. Originally the Old Town lay some six to nine feet below the modern street level. The area was, however, subject to repeated flooding, which is why the street level has been raised bit by bit since the late 13th century. Many houses now have Romanesque rooms hidden in their basements. The historic core of the Old Town is built on these foundations. Every age has left its signs for us to read. The overwhelming influence of the Baroque cannot be overlooked. However, it only finds its expression in individual buildings, it has not changed the structure of the district. The only large intrusion is the massive building of the Jesuit College, the Clementinum. Here and there you can also see traces of the 19th and 20th centuries, for the development of the river bank gave the big city a chance to break in. However, apart from demolition of the Jewish quarter, the entire district has hardly lost any of its charm.

The present-day appearance of the streets is still marked by the unceasing succession of houses with the most varied facades. This is a living district with a well-balanced mixture of homes, offices, shops, small businesses, several schools and leisure facilities, all of which play a part in forming the impression given by the district.

The first settlements on the site of the Old Town for which there is any historical evidence date from the 10th century. They concentrated around the crossing of three important trade routes, which met at the ford of the Vltava, a little downstream from where the Charles Bridge stands today. According to a contemporary report, a large market place with numerous stone houses covered the site of the present Old Town Square. As the years went by this market place grew, and was fortified with a city wall in the early 13th century. Around 1230, the settlement received city rights. By this time it was possible to speak of a large town, in European terms. In 1338, the citizens of the Old Town received the right to their own Town Hall, and in the years that

Preceding pages: Pilsner, Budweiser. Below, the Jan Hus memorial.

followed, during the rule of Charles IV, the city experienced an immense economic and cultural boom.

The Charles University, the oldest university in Central Europe, was founded in 1348. Even if the importance of the imperial residence did diminish later on, the Old Town kept its leading position in Prague. When the four independent cities became one administrative unit in 1748, it was the Town Hall in the Old Town that became the seat of the administration.

Old Town Square

All the busy streets near the border of the New Town lead the visitor who crosses over into the Old Town to the Old Town Square (Staroměstské náměsti). The streets lead to the square from all sides like the rays of the sun and make it a natural center.

The **memorial** in the middle of the square honors the great reformer Jan Hus and was put up on the 500th anniversary of his death (Jul. 6, 1915). In recent years the Old Town Square has also been the scene of large gatherings.

The houses on the east side of the square form a singular backdrop. This juxtaposition of the most varied building styles is typical for the Old Town and, together with the towers of the Teyn church, it gives the Old Town Square its special character. To the left you can see the **Palais Kinsky** with its Late Baroque facade, which already incorporates some Rococo elements. It was built by A. Lurago, following plans by Kilian Dietzenhofer. Today you will find the National Gallery's collection of graphic art here. To the right of the palais is the Gothic house **The Bell** (*Dum u zvonu*), which has recently been restored and had its original facade replaced. Various exhibitions are held in the interior, which is worth

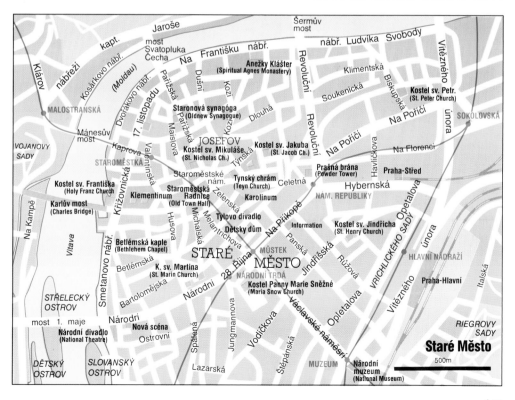

Staré Město

seeing. The two neighboring houses are connected by an arcaded passage with ribbed vaulting. To the left, the **Teyn school**, originally a Gothic building, was rebuilt in the style of the Venetian Renaissance. Attached on the right is the early Neo-classical house **The White Unicorn**.

You can gain access to the **Teyn church** through the Teyn school or from Celetná ul. no. 5. It was the third church built (in 1365) on this site, the successor to a Romanesque and an Early Gothic building. Up until 1621 it was the main church of the Hussites. The tall nave received Baroque vaulting after a fire. The paintings on the high altar and on the side altars are by K. Skreta, the founder of Bohemian Baroque painting. Other remarkable works of art are the Gothic Madonna (north aisle), the Gothic pulpit and the oldest remaining font in Prague (1414). To the right of the high altar is the

tombstone of the Danish astronomer Tycho Brahe (1546-1601) who worked at the court of Rudolf II. The window immediately to the right of the south portal is a curiosity—through it you can see into the church from the neighboring house. One of those who could by this means experience the services without leaving his apartment was, for a time, the author Franz Kafka.

The Baroque **St. Nicholas Church** (Kostel sv. Mikuláše) on the other side of the square is also the work of Kilian Dietzenhofer. The statues on the facade are by A. Braun, a nephew of M.B. Braun. The unusual proportions of the church have come about because houses originally stood in front of the building, completely separating it from the square. It is interesting to see how the architect has succeeded in creating so perfect a building in such a relatively small space. The house to the left of St. Nicholas Church, by the way, is the birthplace of Franz Kafka.

Old Town Hall

The area of the small park opposite the church was in earlier years occupied by a Neo-gothic wing of the **Old Town Hall** *(Staroměstská radnice)*. It was destroyed in the last days of World War II. If you walk around the Town Hall Tower, which protrudes into the Old Town Square, you gain an unobstructed view of the historic part of the Town Hall. At first, the house next to the tower on the left was purchased by the citizens of the Old Town and declared a Town Hall. Later, three further houses in this row were bought. The tower was built in 1364 and later had the oriel chapel added. The **astronomical clock** on the tower dates in its earliest form from 1410. It consists of three parts. In the middle is the actual clock, which also shows the movement of the sun and moon through the Zodiac. The representation is in accordance with the

Left, the Powder Tower. Right, the famous Prague Apostle

geocentric views of the time. Underneath is the calendar, with signs of the Zodiac and scenes from country life, symbolizing the 12 months. The artistic work on the calendar is by the well-known Czech painter Josef Manes.

The upper part is a popular attraction. Every day on the hour the figures play the same scene: Death rings the death knell and turns an hourglass upside down. The 12 apostles proceed along the little windows, and a cockerel flaps its wings and crows. The hour strikes. To the right of Death a Turk wags his head. The two figures on the left are allegories of greed and vanity.

Also on the hour a guided tour leaves for a trip through the historic rooms of the Town Hall, which also contains exhibition rooms.

The memorial tablets on the Town Hall Tower are reminders of some of the important events that have taken place in this square: the execution of the radical Hussite preacher Jan Zelivsky (1422), who became known to history through the First Defenestration of Prague; the executions of the "27 Bohemian gentlemen" (1621), a punishment of the leaders of the rebellion of 1618, intended to serve as an example to others; the liberation of Prague by the Red Army on May 9, 1945.

Side Streets

Passing the house **The Minute**, you come to the **Small Square** (*Malé náměstí*). Here you will find a fountain with a remarkable Renaissance grille. Apart from a few house signs the facade of a traditional ironmonger's, richly painted with ornamental and figurative motifs, is notable. The interior design of the Square has kept the atmosphere of an old-fashioned Prague shop. Although the Small Square is only a few steps away from the big neighboring

In the summer people in Prague like to sit outdoors.

square, it has a quite different atmosphere. Already you feel yourself surrounded by the Old Town and its narrow twisting streets.

In order to enjoy this change of atmosphere, you should make a short trip into the neighboring little streets—the Karlova ul, for instance, which bends to the left, and then straight on into the Jilská ul. Soon you will see on your left house no. 18, which in earlier years bore the name **Two Stags with one Head**. An unassuming arch is followed by a passage to the Michalská ul. (the so-called "Iron Gate"). It links up with another passage through a palace courtyard with Renaissance arcades. However, there is also another route on the left, which crosses the courtyard of a monastery containing **St. Michael's Church**, in which Jan Hus preached. Both ways meet up again in the Melantrichova ul. just before the Kožná ul. leads into it. The first house on the left,

The Two Golden Bears, is a beautiful example of Renaissance architecture. The Kozná ul. leads you out of the labyrinth back to the Old Town Square.

Celetná ul. and Powder Tower

The Celetná ul. is named after the medieval bakers (calty) of small loaves. It is one of the oldest streets in the whole of Prague, for its course follows the line of the old trade route to the east where it left the Old Town markets.

Following the large-scale restoration undertaken in recent years, most of the Baroque facades in this model street shine in new glory. Of particular architectural interest is the Late Baroque **Palais Hrzán** (no. 12). Nearby is the wine bar **The Golden Stag** (*U zlatého jelena*), which is situated in what were originally the rooms of one of the oldest stone houses in Prague. An architectural rarity of quite a different order is

Wedding coach in front of the Old Town Hall.

the unique Cubist house **The Black Mother of God** (no. 34), designed by Joseph Gocar (1911-12).

At the end of the Celetná ul. is the Late Gothic Powder Tower (Prasná Brána). It was built in the second half of the 15th century as an impressive city gate, replacing an older gate which had stood on this site before. Its special status among the 13 gates of the Old Town fortifications came about because the Royal Court (no longer in existence), which acted as the royal residence in the 15th century, was right next door. It received its name in later years, when it was used as a gunpowder store. The Neo-gothic roof and walkway were added during rebuilding in the second half of the 19th century. The tower is open to the public on Wednesdays and weekends from April to October and offers an interesting view.

On the site of the Royal Court mentioned above, the **People's House** *(Obecní dum)* was built in the years 1906-11. The splendid art nouveau building was created in response to the politically and economically strengthened national consciousness of the Cezch bourgeoisie around the turn of the century. A whole generation of Czech artists worked on this building. Appropriately enough, it was in this building that the Czech republic was declared in 1918, after World War I. The interior, which is still in its original condition, can be seen in the restaurant and the café belonging to the house. There are also various rooms which are used for social events and the **Smetana Hall**, a famous concert hall.

Another example of Prague late art nouveau (similar to the Viennese Secessionist style) is the **Hotel Paris**, which lies next door to the People's House of 1906.

If you enter the little alleys at the back of these buildings, you soon come to **St.**

The Towers of the Teyn church.

Jacob's Church *(Kostel sv Jakuba)* in the Malá Šupartská ul. You can also get to it from the Celetná ul., through one of two passages in the houses no. 17 and no. 25. Like so many churches in Prague, St. Jacob's (founded 1232) was rebuilt several times until it attained its present Baroque form. Notable works of art are the reliefs on the main portal, the ceiling frescos and the painting by V. V. Reiner on the high altar. Particularly valuable from an artistic point of view is the tomb of Count Vratislav Mitrovic, the work of J. B. Fischer von Erlach and F.M. Brokoff. Organ concerts are regularly held in the church because of the excellent acoustics.

The cloisters of the former Minorite monastery adjoin the north side of the church. Musical instruments of all kinds can be heard in the former monks' cells in the upper storey, for the monastery is now a music school.

Between St. Jacob's and the Teyn

The most beautiful view of the Melantrichova can be seen from the Old Town Hall.

church lies the Teyn Court, also known as the *Tyn*. It is a quiet place, separated from the rest of the city, with plenty of atmosphere. Originally it offered protection to foreign merchants. The whole complex, whose origins go back to the 11th century, is currently being renovated. You can use the Tynská ul. to get back to the Old Town Square. It leads around the Teyn Court to the right and leads you to the north portal of the Teyn church. The covered end of the street, the Teyn Court gateway and the church portal with the magnificent tympanum from Peter Parler's workshop together make up one of the most picturesque corners of the Old Town.

Pařížská to Crusader Square

The impressively proportioned Pařížská třída (Paris Street) leaves the Old Town Square by the St. Nicholas Church and leads to the sights of the former Jewish quarter (Josefov). You should combine a tour of this Jewish district with a visit to the Agnes convent *(Anežsky klášter)*, which lies a little out of the way in the Anezská ul. The convent is the first Early Gothic building in Prague (founded 1234). The whole complex, which included two convents and several churches, fell into decay over the years and parts of it were completely destroyed. After painstaking work lasting many years, restorers succeeded in bringing some rooms back to their original state. These were linked up to form the present-day historic complex by means of carefully reconstructed additions. The convent contains an exhibition of the Craft Museum (19th century crafts) and a collection from the National Gallery (Czech painting of the 19th century).

Leaving Josefov via the street Ul. 17 Listopadu, you will have only a few steps to go before coming to the **Craft Museum** (Uměleckoprumyslové muzeum), which has a public reference library and an exhibition room displaying a variety of objects from various crafts and skills.

Located diagonally opposite the Craft museum is the Rudolfinum (also known as *Dum umělcu* or House of Artists), an impressive building in Neo-renaissance style, which faces the square, *nám. Krasnoarmějcu*. Nowadays it is the seat of the Czech Philharmonia and boasts a magnificent concert hall. From 1919-39 it served as a parliament building.

We recommend that you cross the square now and enjoy the view of the Charles Bridge, the Malá Strana and the castle from the banks of the Vltava. If you carry on upstream along the embankment and then follow the Křižovnická ul. you will pass the massive and somewhat gloomy facade of the **Clementinum** and come to the **Crusader Knights Square** (*Křižovnické nám.*) with its monument to Charles IV.

Right on the first pillar of the **Charles Bridge** is the **Old Town Bridge Tower**, built, like the bridge itself, by Peter Parler. The remarkable statues which ornament the tower are also from the great master's workshop.

The Baroque **church of the Crusader Knights** (*Kostel sv. Františka Serafinského*) on the river bank side of the square is dedicated to St. Franciscus Seraphicus. In the earlier years, the church belonged to the monastery of the "Order of Crusader Knights with the Red Star", the sole Bohemian knightly order at the time of the Crusades. The magnificent cupola of the church is decorated with the fresco *The Last Judgment* by V. V. Reiner.

The houses built into the river were originally the Old Town water mills. The last house, which is decorated with sgraffito and is a former waterworks, now contains the **Smetana Museum**.

Opposite the Bridge Tower you can see the Baroque facade of **St. Savior's Church** (*Kostel sv. Salvátora*), which is part of the Jesuit college of the **Clementinum**. This broad complex was founded by the Jesuits, who were called to Prague in 1556 to assist the return of the country to the Catholic fold. Nowadays the State Library of the CSSR is based here.

Charles Alley

The narrow and twisting **Charles Alley** (*Karlova ul.*) has always been the link between the Charles Bridge and the Old Town Square. In house no. 4 the astronomer Johannes Keppler lived for a while. A little further on, in house no. 18 **The Golden Serpent**, the Armenian Gorgos Hatalah Damashki opened the first Prague café in 1714.

Leave the Karlova, following the outer wall of the Clementinum, and you have only a few steps to go to the Nám. primátora dr. V. Vacka. The art nouveau building in this square is the **New**

Left, friendly young man in uniform. Right, everyone is waiting for the cockerel to crow.

Town Hall with the residence of the Mayor of Prague. The **Palais Clam-Gallas** lies at the corner of the Husova ul. It is a magnificent Baroque building, designed by the Viennese court architect J.B. Fischer von Erlach. The portal ornamentation is by Matthias B. Braun. Opposite the Palais is the **City Library**, which also houses a department of the National Gallery (Czech art of the 20th century).

To the south of the Karlova ul. a network of tiny streets spreads out. They invite you to stroll along them.

Follow the Husova ul. for now. On the right you will notice a facade built in the style of the Venetian Renaissance. This house (no. 19) contains the **Central Bohemian Gallery**, which holds exhibitions of regional art. A little further to the left is **St. Aegidius' Church** *(Kostel sv. Jiljí)*. You get the best impression of the clear, Gothic lines of exterior of the church, which contrasts with the overloaded Baroque interior, if you simply walk around it. The central courtyard of the monastery that adjoins the church to the left has an atmosphere of quiet solitude.

A road runs off at right angles to your right, and leads to the Retě zová ul. House no. 3, *Dum pánu z Kunstatu*, should be mentioned. In its basement the whole ground floor of a Romanesque palace has been preserved. These rooms are open to the public and serve as exhibition rooms for the "Prague Center for the care of Monuments". Further on in the same direction is the little square *Anenské nám.*, which has a cosy atmosphere.

An important memorial to the Hussite past is the Bethlehem Chapel *(Betlémská kaple)* in the square of the same name *(Betlémské nám.)*. Today's building is a faithful reconstruction of the original chapel founded in 1391. Here the mass was said in Czech. The

Christmas tree in the Old Town Square.

plain interior had the pulpit as its focal point and not the altar. It could hold up to 3,000 people. In the early 15th century, the famous reformer Jan Hus preached and worked here. His ideas spread out from this place to all over the country. In 1521, the leader of the German peasants' revolt, Thomas Munzer, also preached in this church.

A picturesque courtyard located right on the western side of the square contains the **Ethnological Museum** (*Náprstkovo muzeum*) with an exhibition of artifacts from Asian, African and American cultures.

Just a little out of the way, on the corner of Ul. Karolíny Světlé and Konviktská ul., lies the **Holy Cross Rotunda** (*Rotunda sv. Kříže*), a Romanesque round church dating from the beginning of the 12th century.

"Every Bohemian is a musician".

A curious church building awaits you in the Martinská ul. Originally Romanesque, later rebuilt in Gothic style, the church of **St. Martin in the Walls** was incorporated into the city walls. Here, in 1414, holy communion was first given "in both kinds", (i.e both bread and wine given to the laity).

Our tour of the Old Town ends back in the neighborhood of the Old Town Square. The Martinská ul. leads into the **Coal Market** (*Uhelny trh*), and some old market streets follow. Worth taking a look is the picturesque alley *V kotcích*, in which time seems to have stood still. Between the **Fruit Market** (*Ovocny trh*) and the Zelezna ul. lies the **Tyl Theater**. It was opened in 1783 as the Nostitz Theater and is the oldest theater building in Prague. The theater played a large part in the cultural life of the city. To the left of it lies the **Carolinum**, a historic building which belongs to the Charles University. The magnificent oriel in the outer wall is a remnant of the original Gothic building of the 14th century.

Der Anfang des dreifsigjährigen Krieges.

DEFENESTRATIONS OF PRAGUE

March 10, 1948, early morning. The caretaker in charge of heating the Czernin Palace, Karel Maxbauer, found his employer, the foreign minister of the Czech republic, Jan Masaryk (aged 63), dead in the courtyard. Masaryk, son of Tomas Masaryk, the founder of the Czech republic, was the only non-communist cabinet member left after the "bloodless" coup in February 1948. A month ago, 12 non-communist ministers in the coalition cabinet of the communist premier Clement Gottwald had resigned. Under pressure from communist militia and the threat that Soviet troops might invade, state president Beneš appointed a communist cabinet. The sole exception: Jan Masaryk. Was the fall from the bathroom window, 45 feet (15 meters) up, suicide or not?

May 23, 1618. Enraged Protestant citizens of Prague threw three Catholic councillors out of the window of the Hradčany castle into the moat. This confrontation between Catholics and Protestants led to the Thirty Years' War. In order to consolidate his power in Bohemia, Ferdinand II had originally agreed to the terms of Rudolf II's "Letter of Majesty" which guaranteed religious freedom and the unrestricted building of churches. But Ferdinand II soon showed himself to be a supporter of the Counter-Reformation. His harsh and merciless actions led to open revolt. Protestant churches in Bohemia were closed, and some were even torn down. The "Defenestration" by the enraged Estates of Prague had far-reaching consequences. The monks of the Strahov monastery and archbishop were exiled, Ferdinand II was declared deposed in 1619 by the Bohemian Estates. In 1620, the army of the Estates, dependent on foreign help and led by Frederick V of the Palatinate, was defeated by the imperial army in the Battle of the White Mountain near Prague. The power of the Habsburgs and the Catholic church was re-established. In the subsequent "Bloody Trial" in Prague, 22 Czech and five German noblemen were publicly tortured and executed in the Old Town Square.

July 30, 1419. A stone was thrown from the window of the New Town Hall at a procession of armed Hussites. The unavoidable, long-smouldering conflict finally broke out. The Hussites stormed the Town Hall and threw three consuls and seven citizens out of the windows. The Hussite Wars had begun. Jan Hus was born in 1369 in Husinec, Bohemia. He became a priest in 1400, in 1402 he became Rector of the Bethlehem Chapel, and in 1409, Hus was made Rector of Charles University. His fiery speeches attacked the worldliness of the

Catholic church, the immoral behavior and lack of sanctity of the clerics, and promoted a Czech national movement.

Hus' ideas were popular with many of the people and with King Wenceslas IV. He was forbidden to preach by the Archbishop of Prague, and excommunicated in 1411 when he refused to obey. Hus agreed to attend the Council of Constance if granted a safe conduct, and obtained one from Wenceslas' successor Sigismund. However, he was arrested and imprisoned, charged with heresy and burned in 1415.

Left, portrayal of the Defenestration of Prague, May 23, 1618. Above, Jan Masaryk is dead.

NOVÉ MESTO—THE NEW TOWN

Nové Mesto is the New Town district of Prague, and has not nearly as many interesting sights to offer as the other districts described above. However, you should not miss taking a stroll through this district, even if you do have to cover considerable distances on foot. This is one district where you will get a good impression of everyday life in Prague. The shops do not as yet cater for tourists, and the condition of most of the houses proves that more than the Old Town Square is in need of renovation.

Around Wenceslas Square

The gently rising, gigantic former Horse Market is crowned by the martial-looking equestrian **statue of St. Wenceslas**, finally erected by Josef Myslbek in 1912 after thirty years of plans and designs. The people of Prague congregate when necessary in Wenceslas' shadow, as well as in that of Jan Hus in the Old Town. Announcements are made to the crowd, and demonstrations start here. This happened both in 1919 and in 1939, and also during the Prague Spring in 1968. Pictures of the Square at that time went all round the world.

The measurements of Wenceslas Square (*Václavské nám.*) are overwhelming. 179 feet (60 meters) wide and 2,230 feet (680 meters) long, its size is well in advance of its time. Nowadays the impression given by Wenceslas Square is of hotels, pubs, restaurants, cafés, banks and department stores. It is a gigantic and busy boulevard, along which half the population of Prague seems to stroll in its leisure time. What you can't find in Wenceslas Square won't be found anywhere else in Czechoslovakia.

The old two-storey Baroque houses that once surrounded the square have gone. They have been replaced by six and seven-storey buildings, of which only a few, such as the Hotel Europa, still retain their art nouveau facades. Behind the statue of St. Wenceslas, so redolent of history, the square is closed by the National Museum (*Národní muzeum*). A successor to the National Theater in the Národní tr., it was built in 1885-1890 by the Prague architect Josef Schulz. Although he was assistant to Zitek, the architect who designed the National Theater, Schulz wasn't quite a match for Zitek with this copy. The building that was intended to become the spiritual and intellectual center of the Czech nation seems rather unfortunate and clumsy.

Next door, replacing the old Produce Exchange, is a conspicuous glass building, the **New Parliament Building.** Almost in the shadow of the big Parliament Building lies Vítezného

Preceding pages: view from the National Museum across Wenceslas Square. Below, equestrian statue of St. Wenceslas.

února 8, the **Smetana Theater** *(Smetanovo divadlo)*. It was built in 1888, a successor to the wooden "New Town Theater" which had stood on the same site. At that particular time it was the "New German Theater", the second largest German language stage in Prague and for that reason, it is not entirely without its problems.

If you walk down Wenceslas Square, the main shopping area of Prague begins about on the level of the streets Jindrísská and Vodickova. Most of the expensive shops, department stores and bookshops can be found around the Metro station Mustek, in the pedestrian precinct Na přikopě and the 28. rijna.

At the lower end of the square two streets (both for pedestrians) branch off. The name of the Metro station *Mustek* is a reminder that a bridge which led to the Old Town once stood on this site. Remains of the bridge can be seen in the underground station. The pedestrian

precinct of Na přikopě (By the Moat) follows the course of the old fortifications towards the Powder Tower. Don't waste too much time on the less than interesting displays in the shops. More interesting are no. 12, the **Palais Sylva-Tarouca** (built in 1670 and extensively altered in 1748), and no. 22, which dates from the 18th century and is now the **Slavonic House**, formerly the Palais Prichovsky, then the **German House**.

Right opposite is the **People's House** *(Obecní dum)*, built in a Seccessionist style. The passage from Na přikopě 11, next to the Café Savarin, and to Wenceslas Square is also attractive. Panská ul., in which the accommo-dation office of **Cedok** is situated, leads off from the Na přikopě.

On the other side, the street 28. rijna leads to the Námestí Jungmannova with its memorial to the Czech linguist Josef Jungmann (1773-1847). Here, many

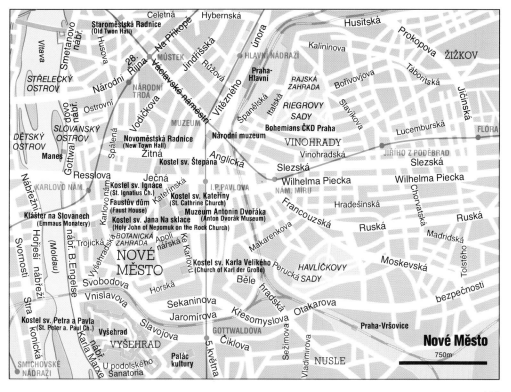

133

visitors to Prague go straight past the gate of the Franciscan rectory and overlook the church of **St. Mary of the Snows** *(Kostel Panny Marie Snezné)*, which was planned to be a massive building. Today all you can see of the church which was founded by Charles IV in 1347 as a coronation church is the choir. The plans envisaged a church comparable to St. Vitus' Cathedral, a three-aisled Gothic cathedral church, and to become the tallest building in Prague. However, shortage of money and the start of the Hussite wars saw to it that the plans were never fulfilled. This is why the proportions of the church look rather odd. Inside, the 16th century altar and the font dating from 1459 are worth special attention. You can get a good view of the church from the little park behind the dum sportu, which leads to the Alfa passage and so on to Wenceslas Square.

The Národní trída leads off the Jungmannova nám. and on to the Vltava and the 1st May Bridge. Notable are the glass facade of the **New Theater** *(Nová scéna)* and the **National Theater** *(Národní divadlo)*. Before you get to the National Theater, and just before the New Theater, you will pass the **Ursuline Convent** and the **Ursuline Church**. The church is currently being restored, and in front of it is a group of statues, and notably among them, is St. John of Nepomuk with cherubs, by Ignaz Platzer and dating from 1746-47. Nowadays the former convent buildings contain an excellent wine bar.

The *Národní divadlo* is, purely and simply, the symbol of the Czech nation. In 1845 the Estates, with their German majority, turned down the request for permission for a Czech theater. Money was collected, and the building of a Czech theater declared a national duty. In 1852 the site was bought, and the

The National Theater is mirrored in the glass of the New Theater.

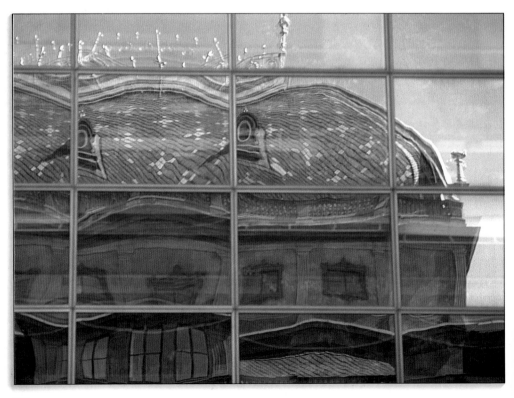

foundation stone was laid in 1868. The building, built by Josef Zitek in a style reminiscent of the Italian Renaissance, was opened in 1881 with a performance of Smetana's *Libussa*.

In August 1881, only two months after the opening, the National Theater was burned down. Under Josef Schulz's direction, it was rebuilt and re-opened in 1883.

Right opposite the National Theater, in the former **Palais Lazansky**, is the **Slavia**, one of the last great coffee houses of Prague.

Enroute to Charles Square

Take a look from the 1st May Bridge, which as the second oldest bridge in Prague has often had to change its name, up the Vltava, and you will see the Slav island *(Slovansky ostrov)* with the Sophia Hall and the Café Mánes. At its upper end, near the Jiráskuv Bridge, is the end of the Resslova, in which lies the Orthodox church of **St. Cyril and Methodus** *(Kostel sv. Cyrila a Metodeje)*. The starting point for boat tours on the Vltava lies between the **Jiráskuv** and **Palackého** Bridges. You can get into the (usually closed) church through the former sacristy. After the assassination of the "Reich Protector" Reinhard Heydrich on May 17, 1942, three of the assassins and four other members of the Resistance barricaded themselves into the crypt. Their hiding place was discovered on the June 18 and they were removed by the SS after a bitter struggle.

Today a number of photographs and documents are displayed in the crypt and it serves as a memorial. In revenge for the assassination, Heydrich's successor Karl Hermann Frank ordered the village of Lidice near Prague to be burned to the ground. This was done on Jun. 10, 1942 and the 199 men in the

Left, the National Theater and the New Theater. Right, in the MAJ department store.

village were shot. The women and children were deported to concentration camps around the country.

If you turn off from the Resslova into the Na Zderaze and then into the Na zborenci, you will come straight to the Kremencova, which boasts one of the most famous of Prague's beer pubs, the **U Fleku**, a brewery and small restaurant which serves a special dark beer—something not to be missed.

U Fleku, together with **The Chalice**, is probably the pub most often visited by tourists in the peak season in Prague. Every room has a different name. For instance, you not only find the **Velky sál** (Great Hall), but also the Jitrnice, which means "liver sausage". The shady beer garden is extremely popular in summer, and in the evenings a small brass band adds to the atmosphere. A traditional Czech cabaret appears every evening in the U Fleku and is one of the last of the Prague cabarets.

Charles Square

Your route through the New Town district of Prague continues from the Resslova to the Karlovo námestí, or **Charles Square**. It was also part of Charles IV's building project for the New Town and was laid out in 1348. Charles Square was the biggest market in the city and was, until 1848, known as the "Cattle Market". Its present appearance is due to 19th century rebuilding. The monuments in the park portray famous Czech scientists, scholars and literary figures.

More interesting than the square is the **Town Hall** of the New Town *(Novomestská radnice)* at the northern end of the square. It was built in several stages between 1348 and 1418, after the founding of the New Town. Alterations to the south wing followed a hundred years later, in 1520, and the tower was rebuilt in 1722. Its present appearance is

A cause of some controversy in Prague—the MAJ department store.

due to reconstruction of the building in 1906, restoring its original appearance. This is where the first Defenestration of Prague took place, initiating the Hussite wars which lasted for 15 years.

In the middle of Charles Square, on the eastern side, is **St. Ignatius' Church** (*Kostel sv. Ignáce*). This is where the Jesuits built their second college in 1659, following the Clementinum. The college was dissolved in 1770 and since then the large complex has served as a hospital. The church, built in 1665-70 by Carlo Lurago, was extended in 1679-99, with a pillared hall and arcade designed by Paul Ignaz Bayer. Inside, among other works of art, is an altarpiece *Christ in Prison* by Karel Skréta (1610-1674).

A walk through the New Town doesn't offer nearly as much variety as a tour of Malá Strana or the Old Town. If you want to see interesting buildings and sights here, you'll need good, durable shoes. The facades of the New Town apartment houses are not really all that exciting. The soot that has been falling for decades onto the city is thick on the walls, and so everything looks a little dilapidated. Large-scale renovations, like those in Charles Alley, for instance, are not in evidence here. Even the present-day pharmacy which achieved fame as the **Faustus House** (*Faustuv dum*), is a sorry sight. In this building, originally a Renaissance house, at the bottom end of Charles Square, the alchemists Edward Kelly and Ferdinand Antonín Mladota conducted their experiments. The latter also entertained his guests with a series of conjuring tricks and a magic lantern. In Prague, a city sensitive to such activities, that provided reason enough to give the house its peculiar name. Kelly, no more a serious scholar than Mladota, was supposed to discover the Philosopher's Stone for Rudolf II, but did not succeed. Rudolf had the Englishman, whose ears had been cut off in his own country as a punishment for fraud, thrown into a cell. The search for the Stone probably lasted too long for Rudolf's liking. Kelly died of poison in prison, after two attempts of escape.

There are fewer legends surrounding the church of **St. Nepomuk on the Rock** (*Kostel sv. Jana Na skalce*) in the Vysehradská, just around the corner. It is a pity that it is so difficult to get into this very beautiful Baroque church. Like many churches in Prague, it is locked for most of the time. In similar sorry case is the former monastery opposite, the **Emmaus monastery** (*Kláster Na Slovanech*). It goes back to a foundation of Charles IV in 1347 and was destroyed by bombs in 1945. To replace the two towers, two sail-shaped buttresses were added to the church in 1967. They can be counted among the rather few examples of originality in modern architecture in Prague. Unfortunately the whole structure is an

Sculpture in front of the Metro station Národní trída.

illusion. The concrete additions of Frantisek Cerny cover an unhappy ruin. In the cloisters right next door are Gothic frescos dating from the 14th century. The Emmaus monastery became famous as a medieval scriptorium producing Slavonic manuscripts. Today it is the home of a scientific institute, and the frescos can only be seen during office hours.

Only a few steps away from the Emmaus monastery lies the **Botanical Gardens** with their beautiful greenhouse, which is unfortunately also in need of renovation.

Around New Town

Keep following the *Vysehradská*, which, once past the Botanical Gardens, turns into the Na slupi, and you will come to another important (but most often locked) church, *Maria Na slupi*. This former convent church of the nuns of the Elizabethan order is also Kilian Ignaz Dietzenhofer's work, and a rare example of a Gothic church supported by a central pillar.

If you go up the steps of the Albertov, you will come to another building that's well worth seeing, **Charles Church** *(Kostel sv. Karla Velikého)*. The former Augustinian monastery is surrounded by University buildings, and you enter it through a plain gate. The Charles Church itself is obviously an unusual building, as can be seen just from the exterior. The building has an octagonal ground plan and a central dome. It was founded in 1350 by Charles IV and dedicated to Charlemagne in 1377, and is reminiscent of the imposing imperial chapel in Aachen.

Charles Church lies right on the edge of the descent into the Nusle valley, and only a few yards away from its surrounding wall the **Clement Gottwald Bridge** *(most Klementa*

Left, a short spell of sunbathing during lunch break. Right, scene in the U Fleku.

138

Gottwald) arches across the valley. The bridge leads across the valley, over the apartment houses of Nusle that lie beneath it, and its 1,640 feet (500 meters) make it the second longest bridge in Prague. There are two modern glass "palaces" at the end of the bridge. One is the new **Hotel Forum**, completed in 1988, and on the other side is the **Palace of Culture** *(Palác kultury)*, one of the "most notable of modern buildings", officially opened on the Apr. 2, 1981.

If you walk slowly back down the Ke Karlovu and past the Charles Church, you will see on the right-hand side the **Villa Amerika**, the Antonín Dvorák museum. This small, charming building was constructed in 1717-20 as a summer palace for the Michna family. Amidst the rather dreary university buildings the little Baroque villa, named after a former 19th-century pub, provides a pleasant change for the eyes.

The **Lapidarium** opposite, in the church of **St. Catherine**, which can be reached by following the Katerinska, is, however, less attractive. The church also goes back to a foundation made by Charles IV in 1355. It was largely destroyed in the Hussite wars. The octagonal tower is all that remains of the building from that time. The present-day building is the result of rebuilding work undertaken in 1737. The collection of cast figures and woodcarvings cannot be seen at the present time.

Right next to the Villa Amerika and the Lapidarium, in the Na bojisti 14, is one of the most important pubs in Prague, **U Kalicha**. Although it lies in an unassuming street in the New Town district of Prague, the buses parked in front of it are a sign that something special is going on here. And it is, too.

This is where the Prague author Jaroslav Hasek frequently used to

U Fleku— an absolute must for those who admire Czech beer.

drink, and his novel *The Adventures of the Good Soldier Schwejk* has made the "Chalice" famous. "When the war's over, come and visit me. You'll find me in the Chalice every evening at six", Schwejk says to his friend. Today, the pub has become a place of pilgrimage. The walls are covered with paintings and quotes, the waiters dressed up—all is familiar from the book and the film. However, there is one drawback to U Kalicha—the countless coachloads of tourists who arrive at lunchtime. If, by the way, you want to continue on the trail of the Bohemian (in both senses) Jaroslav Hasek, you can visit the house where he was born. It is also in the New Town, on the Skolská which branches off from the Vodickova.

Another famous author who lived in the New Town is Franz Werfel, who lived near the City Park and whose father owned a glove factory in the Opletalova behind the Hotel Esplanade.

The famous **Café Arco** was the meeting place of the "Arconauts"—Franz Werfel, the "rushing reporter" Egon Erwin Kisch, Franz Kafka, Max Brod and the editors of the German language paper *Prager Zeitung*. It lies in the northern half of the New Town, in the Hybernská 16. Werfel, Kisch, Kafka and Brod once dominated the intellectual life of Prague, but those times are now finally over and the Café Arco is no more than a sad relict. Franz Kafka's grave is in the **New Jewish Cemetery** *(Zidovskéhrbitovy)*, and a visit can be easily combined with a tour of the New Town. You get out at the Metro station Line A, Zelivského. The route to the grave is marked on a tablet near the entrance to the cemetery. However, the cemetery is closed to visitors on Saturdays.

The **main railway station** *(Hlavní nádrazí)*. is worth taking a look. It is originally an art nouveau building, with a large and spacious roofed entrance hall. The combination has on the whole been a success. Not far from the station, but already in the district of **Vinohrady** (Vineyards), is the Riegrovy sady, a large and well laid out park along the slopes of the hill. This offers a very beautiful view across the whole city right up to the Hradčany. A big beer garden also belongs to the park. It is not overflowing with tourists, and on most evenings a small brass band plays there.

A little out of the way, no longer in Nové Mesto but in the district of Zizkov, is the **National Monument** *(Národní památník)*, an immense granite-faced cube with the "Tomb of the Unknown Soldier" and the tombs of state presidents Gottwald and Zaptocky. In front of it is one of the biggest equestrian statues in the world—the monument to the Hussite leader Jan Zizka. Only the Stalin monument, which had dominated the Letna hill until it was demolished in 1963, was bigger.

Left, just like in the Schwejk novel— *Schwejkomania* in the U Kalicha. Right, the National Museum.

BONY A KLID—BONY AND QUIET

"Exchange?" is a question which you will hear at least ten times a day in Prague—in the hotel, when the porter, after a long wait, finally gives you the key and adds "Perhaps, if you haven't already..."; in the lift of the hotel, when people address you "Just in case you..."; in the restaurant, when paying your bill, or even in the tourist buses, which are stopped on the motorway by would-be currency exchangers. In fact, you would always find a good opportunity, if—yes, well, if only. We advise you very strongly not to

take up any offers of black market exchange, however tempting.

Dangers of Currency Dealing

Especially in the main tourist season, the newspapers in Prague are full of stories about black market currency rings whose cover has been blown. And many a tourist has had an unpleasant surprise if he exchanged currency with these people. So hands off—even if they offer you triple and quadruple the official rate of exchange.

The story of such a black market currency dealer is told in the Czech film *bony a klid*.

It is a simple tale of a country boy who comes to Prague and there tries to change currency—on the black market, of course. He needs it in order to buy electrical goods for a small disco. He is cheated, naturally, and palmed off with fake dollars. Of course he doesn't just accept that, he goes to the police to make a report.

Eventually he himself is drawn into the circle of organized currency dealers, stops buses full of Western tourists on the motorway and deals, deals, deals. If any scruples come up, they are buried under piles of money. That's the meaning of the title *bony a klid*—Tuzex-coupons and quiet. Until the day that the gang's cover is blown. "Currency dealing is a sort of hobby", a statement made by one who had been caught.

Dreams of Prosperity

The Tuzex shops sell not only Bohemian glassware, spirits and cigarettes for tourists, but also many things important for everyday life in Czechoslovakia. And in Tuzex shops you can only pay in Western currency. And where is that to come from, if not from a little illegal dealing?

The Tuzex shops are a small-scale paradise—Sony and Johnnie Walker, spare parts for cars and video cassettes, only with much higher prices payable in Western currency. Currency can be exchanged for Tuzex crowns in the State banks and at border crossings.

It's definitely worth visiting a Tuzex shop, and not only because they offer relatively cheap Czechoslovak goods such as Becherovka and Slivovitz, records and glass. A visit to a Tuzex shop gives you an interesting insight into the current material situation in the CSSR, and shows the gulf between what is for many the stuff of dreams and what is actually available.

Preceding pages: the People's House; Pionkrál, mythical kings, in Vojanovy sady. Left, film poster. Right, window cleaner outside the Hotel Forum.

FROM LATERNA MAGICA TO "SPEJBL AND HURVINEK"

Theater in Prague is based on old traditions. The spectrum is broader and enlivened by more original productions than many a "Westerner" imagines. It reaches from Laterna Magica, a "theater between dream and reality", via the the three theaters of the National Theater with their programs of top class opera, plays and ballet, to the operetta productions of the **Hudební divadlo Karlín** theater and the record theater **Hudební divadlo—Lyra Pragensis**. With *Spejbl and Hurvínek* it crosses the boundary

Prague Information Service, published in various languages, where and when such performances take place.

Laterna Magica is known all over the world as a synonym for perfect illusions, for a fascinating, minutely exact mixture of theater, film, mime and dance. All these disparate sections never have an independent effect, they always work as a whole. For 30 years the 60 or so members of the company have tried to make fantasy and poetry, comedy and tragedy real. By means of well-

of puppet theater for children; it includes the mime Ladsilav Fialka, who has been famous for years and appears in the **Divadlo Na zábradlí**, and the younger mime theater **Bránické divadlo pantomimy**, and for a change it offers choice unusual attractions such as the **Komorní opera Praha, Cerné divadlo** (the Black Theater), or the **Prazsky Komorní balet**. In the outskirts of Prague there are little theaters and companies that often spring up quickly and for various reasons disappear just as quickly. Every visitor to Prague can read in the daily national papers or in the monthly program of the

planned technical skill with projections, movable walls, stage props and actors, the plays move between light and shadow, uniting film and theater, mime and dance into one extraordinary stage experience. The visual impression, the dialogue between stage and screen, is at the center of the spectacle. Every new production by this "theater of light" is governed by a new key concept. Following on from well-established programs such as *Magic Circus*, performances such as *Odysseus* (adapted from Homer) or the ballet *Minotauros* with a libretto by the Swiss author Friedrich Dürrenmatt have achieved worldwide recognition.

What do the three theaters of the National Theater—the **National Theater** (*Národní divadlo*), itself the **Smetana Theater** (*Smetanovo divadlo*), and the **New Theater** (*Nová scéna*)—have to offer opera and ballet lovers? The National Theater is the largest theatrical institution in Prague and employs 2000 people. Its program includes not only drama and opera but also works by local composers. The showcase of Prague theater provides good orchestral music, average

world-famous folksy opera *The Bartered Bride (Prodaná nevesta)*; *The Secret (Tajemství)*, a comic opera dealing with a small town feud; *The Devil's Wall (Certova stena)*, another comic opera which describes the quarrel of the devil with a dishonest hermit; the society operetta *Two Widows (Dve vdovy)*; and finally Smetana's opera *The Branderburgers in Bohemia (Branibori v Cechách)*, which had its first success in 1866 shortly before the battle of Sadowa and has been in the repertoire ever since.

singing and ballet without any special qualities, and surprisingly poetic, modern productions directed by J. Svoboda.

The famous Czech composer Smetana is particularly well represented at present. Visitors to Prague can admire the whole range of his romantic operas: *Libussa* (about the mythical princess who founded Prague); the village opera *The Kiss (Hubicka)*; *Dalibor*, set in a world of medieval knights, the

Antonín Dvorák, another distinct Czech composer who received recognition from the musical establishment before Smetana, still has his opera *Rusalka* and other works performed. Martinu has also now become famous, but at present only his opera *Ariadne* is in the repertoire. Admirers of contemporary music will mostly find their tastes represented by Janácek's works. In this case, however, Prague does at least have the privilege of authentic performances, including *Jenufa (Její pastorkyne)*, an opera as tragic as *Katja Kabanova (Kata Kabanová)*, which is based on Ostrovskij's *Storm*; the poetical

Left, between dream and reality—a performance by Laterna Magica. Above, the seats in the National Theater.

Zábradlí

PANTOMIMA
LADISLAVA FIA

and philosophical work that deals specifically with nature and the human race, *The Cunning Little Vixen (Pribehy Lisky Bystrousky)*; and the well-crafted criticism of bourgeois life, *Mr Broucek's Excursions (Vyleti pana Broucka).*

The chamber opera company **Komorní opera Praha** uses younger voices, and performs among other works Pasiello's *Cosi fan tutte*. The company performs in the Palace of Culture or in the *Klicperovo divadlo.*

The *Cerné divadlo,* Jirí Srnec's famous Black Theater, is based in Prague but rarely performs here, as the company is mostly on tour abroad. Equally rare are performances by the **Prague Chamber Ballet**, *Prazky Komorní balet*, directed by P. Smok. The company has no theater of its own and is also on tour abroad more often than it performs in Prague. Performances on record and recitals are held in the small music hall **Hudební divadlo—Lyra Pragensis**.

The **Hudební divadlo Karlín**, which is the **Karlín Music Theater**, performs operettas, often lavish productions including *Die Feldermaus, The Land of Smiles* and even *My Fair Lady.*

The theater **Divadlo Na zábradlí** is almost besieged by enthusiastic visitors whenever the mime artist Ladislav Fialka performs. This artist is a follower of the old French tradition of mime and over the last few decades he has evolved his own Prague style. His performances remain firmly within the tradition of "silent" mime, and are composed of individual scenes within the framework of a single concept.

Amusing incidents are described in the *Etudes*, and *The Loves* deals with all forms of human love from youth up to very old age. *Sny* comprises scenes accompanied by classical Czech music, *Noss* is a moving modern drama about human dignity. Fialka's own homage to the great mime Jean Gaspard Debureau, *Funambules*, has an effective beginning: the mime suddenly rushes on

stage, followed by the sound of grenade explosions and the rat-a-tat of machine gun fire. The noise dies down, and Fialka begins to project the personalities of of the great comedians, Chaplin, Grock, Lloyd and Debureau. A small curtain, blazing with the colors of earlier cabarets, falls, and the performance continues. Fialka created his mimes with the aid of remaining scripts, contemporary reviews and reminiscences of Debureau's contemporaries.

Mime has developed further in Prague in recent years. In 1981 the *Bránické divadlo pantomimy* was formed, the **Branik Mime**

Theater, in which various groups such as CVOC or MIM-TRIO appear.

Spejbl and Hurvínek is a **puppet theater**, a perfect entertainment choice, for children as well as adults. The theater has become well-known in the West through TV performances. It is the only theater company with two main comic figures: the narrow-minded father Spejbl, "born" in 1920, and his son Hurvínek, some six years "younger", who is more exuberant, but also more intelligent. If you happen to be visiting Prague with children, don't miss an opportunity to see a performance.

Left, Ladislav Fialka is already a classic performer of mime. Above, in front of the Novo divadlo—the New Theater.

THE FORMER PRAGUE GHETTO

The former Ghetto was renamed "Josephstown" *(Josefov)* during the Age of Enlightenment, in honor of the reforming Habsburg emperor Joseph II. Later it became the fifth district of Prague. After the clearance program of 1890-1910, decided upon and carried out by the Prague city administration, only the Jewish Town Hall, six synagogues and the old Jewish cemetery remained. These are now administered by the State Jewish Museum. The National Socialists wanted to turn it into an "exotic museum of an extinct race", but after the liberation, it became the home of the largest collection of sacred Jewish artifacts in Europe.

The **Jewish Town Hall**, Maiselova 18, was constructed in 1586 in a

Renaissance style by Pankratius Roder. The alterations in 1756 were the work of Josef Schlesinger. From 1900-1910 the southern part was added.

The **Old New Synagogue**, the oldest remaining synagogue in Europe, is an unparalleled example of a medieval two-aisled synagogue. Nowadays services are still held here. On the outside the building has a plain, rectangular shape, a high saddle roof and a Late Gothic brick gable. The outer walls with their narrow pointed windows are strengthened by buttresses. The low annexes surrounding the main building served as entrance hall to the synagogue and as the women's aisle. The consoles, the capitals of the pillars and the vaulting are all richly decorated with relief ornamentation and plant motifs.

In the center of the main aisle, between the two pillars, is the Almemor with its lectern for reading the Tora, separated from the rest of the interior by a Gothic screen, decorated with asses' back motifs. In the middle of the east wall is the Tora shrine, formed of two Renaissance pillars on consoles, with a triangular tympanum.

The **High Synagogue** was originally part of the Jewish Town Hall built by Pankratius Roder, and was a functional part of it. However, in 1883, the entrance to the Town Hall was closed, which also include an interior staircase. The hall of the synagogue is almost square, lit by numerous high windows, and gives a very worldly impression. The walls in the lower room are divided into three by flat pilasters, a device which echoes the crescent vaulting and the arrangement of the windows in the north wall. The central vaulting with its rich stucco decoration, which echoes the profile effect of Gothic rib vaulting, shows how Renaissance forms adapted to Late Gothic taste. In the lower room there is also a permanent exhibition of sacred textiles.

Left, the pinnacles of the Old New Synagogue. Right, the Jewish Town Hall in the Maiselova.

The **Maisel Synagogue**, founded by Mordecai Maisel, leader of the community in the old Jewish town, as a family synagogue in 1590-1592, was built by Joseph Wahl and Judah Goldschmied in a Renaissance style, altered to a Neo-gothic style in 1893-1905 by Professor A. Grotte. A permanent exhibition of synagogue silver has had its home here since 1965.

The **Spanish Synagogue**, Dusni 12, was designed in 1868 by V.I. Ullmann. It has a square ground plan and a mighty dome covers the central hall. Metal constructions, which open out into the main aisle, lie on three sides. The marvellous decoration of the interior earned this synagogue the name "Spanish Synagogue".

Services were held in the **Cells Synagogue** until 1939. This is a Baroque building with a longish hall and barrel vaulting. In 1694 it was built to replace the little "cells", buildings which served as houses of prayer and classrooms. The building has two rows of round arched windows in the south wall, facing the cemetery. The walls are divided by pilasters which support the roof beams under a pronounced projecting sill. The synagogue is now used to exhibit old Hebrew manuscripts and printed works.

The **Pinkas Synagogue** is architecturally a very beautiful Renaissance building. It came into being in 1535 in a specially adapted private house belonging to the leading ghetto family of Horowitz. In 1625 it was rebuilt in a Late Renaissance style by Judah Goldschmied and extended by the addition of a women's gallery, a vestibule and a meeting hall. Since 1958, the synagogue has been a memorial to the 77,297 Jewish victims of the Holocaust who came from Bohemia and Moravia. The inscriptions on the side walls list name, date of birth

Inside the Old New Synagogue.

and date of deportation to the extermination camps, in alphabetical order, for each victim.

The **Old Jewish Cemetery** is reckoned to be one of the 10 most interesting sights in the world. It came into being in the 15th century, when pieces of land on the northwest edge of the ghetto were bought up. Burials continued here until 1787. The number of graves is much greater than the remaining 12,000 gravestones. Existing graves had to be covered with earth to form new graves, which is what has caused the hilly landscape of the cemetery and the characteristic layering of several graves one above the other.

The inscriptions on the stones give the name of the deceased, the father's name—in the case of women, the husband's name as well—the date of death and the date of the funeral. The majority of the inscriptions consist of poetic texts expressing grief and mourning. The reliefs portray the name of the deceased, the profession, the connections with priestly or Levite families (blessing hands) or membership of the tribe of Israel (grapes).

The oldest monument is the tombstone of the poet Avigdor Karo, dating from 1439. Also buried here in 1609 were the famous scholar and supposed creator of the Golem, Jehuda Löw; in 1601 the leader of the Jewish community, Mordecai Maisel; in 1613 the scholar and astronomer David Gans; in 1655 the scholar Schelomo Delmedigo; in 1735 David Oppenheim. A splendid tomb was built in 1628 by Wallenstein's financier, Jakob Bassevi von Treuenburg, specially for his wife Hendele.

In the Neo-romantic **House of Ceremonies** there is an interesting exhibition of children's paintings and drawings from the concentration camp of Theresienstadt.

The kosher restaurant in the Jewish Town Hall.

FROM THE PRAGUE GHETTO

The Prague Ghetto is one of the oldest in Europe, dating back to the 10th century and possibly earlier. By the 17th century it was a flourishing community and a focal point for Jewish culture in Central Europe.

The Jewish community in Prague was allowed to develop freely after 1848 (when the laws segregating the Jews were finally repealed), until it was almost completely destroyed in the Holocaust following the occupation of Bohemia and Moravia on Mar. 15, 1939 by the Nazis. Today there are some 1,200 members of the Jewish community left. There are also about another 1,500 people of Jewish descent living in Prague.

The Nazis systematically planned and carried out the "Final Solution" to the "Jewish problem". The Jews were first of all forced to register, then excluded from economic activity, then physically separated from the rest of the community. They were marked, insulted, sentenced to perpetual poverty, and evicted from their homes. They were deported to concentration camps, starved and tortured physically and mentally. For those who could no longer work, the death sentence was inescapable.

In the "Protectorate of Bohemia and Moravia" the Jews suffered everything, apart from the "Reichskristallnacht", that had been happening to the Jews in Germany from 1933 on.

Immediately after the occupation of Bohemia and Moravia by Hitler's army, the "Nuremberg Laws", passed in 1935 "to protect the German race", were declared to be in force retrospectively.These laws had deprived German Jews of their citizenship and turned them into "subjects of the state". The same now happened to Czech Jews.One persecution of the Jews followed another—culminating in the deportations of 1941.

The first five transports, each with 1,000 people mainly from the Jewish intelligentsia, i.e. doctors, artists, lawyers, were deported to the so-called Litzmannstadt Ghetto in Lodz, which the Nazis themselves designated a starvation camp.

A month later the old fortress of Theresienstadt in Bohemia was declared a Jewish ghetto, and received further deportations not only from Prague, Bohemia and Moravia, but from all over Europe. Theresienstadt was not itself an extermination camp, but it is from here, that the Jews were sent on for so-called selection and to the gas chambers of Auschwitz.

However, even in times of the greatest debasement of humanity there remained

artists in Prague who had enough courage at least to lessen the impact of the humiliations they suffered. Authors wrote under the names of their friends, cultural afternoons to discuss contemporary poetry were held in private apartments. An amateur drama group led by the famous Shakespeare translator Erik Saudek put on plays. The author Norbert Fryd's verses, *A Horse Decorated with Flowers*, an introduction to the alphabet, were set to music by the composer Karel Reiner and performed by both for the children of the orphanage. In the orphanage in Kosire Glucks, the opera *The Deceived Kadi* was also performed.

Left, tombstone on the Old Jewish Town Hall. Above, in the concentration camp.

VYŠEHRAD

The Vltava flows down from the Bohemian forest and reaches Prague by the rock of the Vyšehrad. This is where, according to legend, the age of myths ended, and the rule of the wise women, skilled in magic, was replaced by the rule of men. The marriage of the Princess Libussa to the farmer Přemysl brought this change about, and their successors ruled over the weal of the Czech people up until 1306 of our era. It was here on this rock, where the couple are supposed to have lived in a magnificent palace, that Libussa had her great vision in which she prophesied the future greatness and glory of the new capital.

Not until the last century did the Vyšehrad enter into the newly revived Czech national consciousness. The legend was embroidered with much imagination and the Vyšehrad became once more the seat of Libussa and the cradle of Czech history. Many artists, poets and painters, musicians and sculptors, historians and architects worked on the site. In this way a sort of memorial was achieved which in its own special way says something about this nation living in the center of Europe—Slavs, surrounded by German tribes, involved in many complex relationships with these neighbors, and yet different in character and speech.

The oldest building on the Vyšehrad is **St. Martin's Rotunda**, a Romanesque church constructed some time after the year 1000. It is one of the oldest Christian churches in the country. There are similar rotundas in various places in modern Prague, for instance the **Holy Cross Rotunda** in the Old Town or **St. Longinus** in the New. They are all that now remains of the cores of former individual settlements.

More is known about the church of **St. Peter and St. Paul**, which the visitor can see in its Neo-gothic form, dating from 1885-1887. At the present time archaeologists are examining the walls of its predecessor on this site. In earlier times the Vyšehrad was the goal of pious pilgrims. Here, in St. Peter and Paul's church, the votive tablet popularly known as the **Madonna of the Rains** was kept. It is now in the collection of the St. George monastery in the Hradčany.

The redundant fortress was demolished in the 19th century, for the Vyšehrad had long lost its strategic importance. A center for the Czech people was created. The heart of these patriotic efforts during the 1870s was the creation of the Slavín, a special cemetery. It was designed by Anton Wiehl and later completed by the addition of the tomb of honor at the end of the main avenue, with its ornamental

Preceding pages: a dog and his master in the Letna Park. Left, Portal of St. Martin's Rotunda.

sculptures by Josef Mauder.

Many of the graves are still regularly decorated with flowers. No "soldiers" or "heroes" are buried here, only poets, musicians and artists. The works of these artists live on in the memory of the nation. The most popular are the two best-known Czech composers, Bedrich Smetana and Antonín Dvorák. Smetana's *Bartered Bride* alone has been performed in Prague more than five thousand times. However, these two composers are not the only musicians buried there. There are also great performers such as violinist Jan Kubelík or the virtuoso Josef Slavík, very famous in his day.

Also buried here is the author of the stories of life in the Malá Strana, Jan Neruda. His stories are of the world of the lower middle classes, living over there in the Malá Strana in the shadow of the palaces. They feature old women and their books of dreams, moonstruck

students, grumbling caretakers, and a number of curious characters from the backyards. Also of Neruda's generation and buried in the Slavín are Svatopluk Cech, Jaroslav Vrchlicky and Karel Hynek Mácha, whose *May Poem* is known to every Bohemian.

The visual artists and painters are represented by Mikolás Ales, Josef Myslbek and Jan Stursa among others, and more modern times by the author Karel Capek, or by Alfons Mucha, who are best known mainly for his art nouveau posters.

The Vyšehrad can easily be reached from the Metro station Gottwaldova, passing the gigantic Palace of Culture. However, you can get to it on foot along a pleasant path from the Slavojova Cikova or directly from the Vltava bank through the thickly wooded park to the castle. From the remains of the old fortifications you can get a beautiful view of the city and the Vltava.

St. Peter and St. Paul on the Vyšehrad.

BOHEMIAN GLASS

Excavations have established that, in the area of present-day Czechoslovakia, glass was known and used in the form of bead necklaces and, later, bracelets. In the early Middle Ages the first blown glass was used for drinking and at the same time windows, glass panes and wall mosaics were manufactured. During the reign of Charles IV, in 1370, the massive mosaic on the south portal of St. Vitus' Cathedral in Prague was made. It shows scenes from the Last Judgment. The splendid windows of St.

in the forests no longer satisfied the refined tastes of the humanist aristocracy who belonged to the court of the first Habsburg rulers. Following Venetian models, thin, refined glass was produced, in harmonious Renaissance styles. Towards the end of the century and after 1600 cylindrical tankards formed the ideal basis for the famous Bohemian enamelled glass.

From 1600-1610, at the Prague court of Rudolf II, the jeweller Caspar Lehmann (1563-1622) of Uelzen had experimented

Bartholomew's Church in Kolín also date from around 1380. Glass for everyday use was, as in other Central European countries, mostly in the form of bottles and phials, greenish or brownish in color.

By the beginning of the 15th century there is evidence for eight glassworks in Bohemia, five in Moravia and another eight in Silesia, which at that time was part of the Bohemian kingdom. The glassworks in Chribská, which still exists today, had already been mentioned in 1427, and had considerable influence on the development of the Bohemian glass industry.

In the 16th century, the rustic glass made

with engraving glass. Lehmann was gem cutter to Rudolf II, and he adapted glass techniques of cutting gems with bronze and copper wheels. This was a new skill used for ornamenting glass, and engraved glass was proudly born.

In Bohemia the production of engraved glass did not begin until around 1680. From this time on glass was exported in great quantities to all sorts of destinations. One example: Between 1682 and 1721 Georg Kreibich from Kamecky Senov made a total of 30 business trips throughout Europe. Among the production managers of famous glassworks in north and northeast Bohemia and in the Bohemian forest the names of aristocratic families such as Kinsky or Harrach can be found.

The classic shapes of the Bohemian Baroque goblet with its polished balustered foot, or the many-sided beaker of fine, thin glass, were intricately engraved with elaborate ornaments featuring flowers, garlands or grotesqueries.

After 1720, the available variety was enriched by the addition of gilded glass and black painted decoration in the style of Ignaz

plain, finely cut and engraved Empire and Biedermeier style glasses.

New success came from 1820 onwards through the thick colored glass discovered by Friedrich Egerman: black hyalith (often with gold Chinoiserie decoration), agate glass, lithyalin (which resembled semi-precious stones) and new uses for ruby glass and opaque white overlay glass, both carved and enamelled, were taken into production. The fame of Bohemian glass was increased in the 19th century by mirror and plate glass

Preissler (1676-1741). Engraved and cut glasses from Bohemia were famous, as was the Bohemian chandelier with pendants of cut crystal glass. Already in the 18th century these chandeliers were being exported to the courts of France and Tzarist Russia.

After some stagnation towards the end of the 18th century, milky Bohemian glass with painted, colorful enamel decoration became tremendously popular. Contrasting trends that developed around 1800 provided

Left, glass window in St. Vitus' Cathedral. Above, the famous Bohemian crystal.

and also by glass coral. The latter survived in Jablonec nad nisou, now under the company name JABLONEX, up until the present day.

Bohemian glass entered the 20th century with names such as Loetz, Lobmayer, Jeykal and others, with imaginative shapes formed by art nouveau, with surprising colors and metallic effects. Apart from its usual domestic and technical functions, modern glass has become an art form. The Arts and Crafts School in Prague produced a particularly excellent generation of artists just after World War II. They have had great success in many important exhibitions in the 1950s and 60s.

LORETO AND NOVY SVET

If you walk up from Hradčany Square toward the Strahov monastery, now the Museum of National Literature, you can hear, every hour on the hour, a delicate tune played by bells as you walk between the palaces.

Many years ago, during the plague, there lived in Prague a mother with her children. One child after another fell sick, and with the last few silver coins that she had left she paid for the church bells to be rung whenever a child died. The poor widow had to give up coin after coin, all her children fell victim to the plague. In the end, after all the children were dead, she herself fell ill and died. However, there was now no-one to have the bells rung for her. Then, all of a sudden, all the bells of Loreto rang out and played the tune of a famous hymn to Mary. So it has remained right up to the present day.

This little unusual story from Prague gives a clear indication of the importance that this shrine has for many people. It is not merely of historic and artistic importance, it is still considered a place of pilgrimage.

The Prague **Loreto** was built as a copy of the famous Lauretanian House in Italy. It stands on the Loretánske nám. and was founded by the Blessed Catherine of Lobkowic, who laid the foundation stone on Jun. 3, 1626. Giovanni Battista Orsi from Como completed the building in 1631. As with so many places of pilgrimage new foundations were added, among them chapels and a large church,

What special quality does Loreto possess, that there are so many imitations of the House of the Holy Family in so many places? In the mid-13th century the armies of Islam

The Casa Santa of Loreto in Prague.

invaded and conquered the Holy Land. At that time two brothers were priors of the Franciscan monasteries in Haifa and Nazareth. When they fled, they probably took all that was most precious to them along. In this way they are supposed to have removed the house of the Holy Family stone by stone, and to have eventually rebuilt it near Renecati, now Loreto in Italy. Later the house, visited by many pilgrims on their way to Rome, was decorated with rich marble reliefs, and the many copies also show this particular ornament.

Just as in Loreto, the outer walls here are decorated with Renaissance reliefs. The interior also strictly follows the Italian model. As a result, you can see in the Prague Loreto a small, bare building which is very probably the copy of a house in Palestine, and in which the Loreto Madonna is honored. This is a slender crowned figure in a long cloak,

Idyllic peace in Novy Svet.

the infant Jesus in her arms.

Kilian Ignaz Dietzenhofer, the Baroque architect of genius from Bavaria, unified the entire complex and surrounded the two courtyards with two-storey walkways. The paintings in these walkways have been heavily over-restored and you can hardly see any traces of their original beauty. However, the wealth of the poetic images in the supplications to Mary is all the more impressive: "Tower of David", "Gate of Heaven", and again and again "Oroduj za nas"—pray for us.

Between the portal and the Casa Santa you can see the Church of the Birth of Christ, a room decorated by notable artists, which was consecrated 111 years after the laying of the foundation stone, almost to the day, on Jun. 7, 1737.

As at many other places of pilgrimage, the pilgrims here have also given votive gifts to the treasury as a

sign of thanksgiving. The gifts of the Bohemian nobility were commissioned from the most notable goldsmiths of the time and are some of the most valuable works of art among liturgical objects in Central Europe. The most remarkable is the **diamond monstrance**, which was a legacy of Ludmilla Eva Franziska of Kolowrat, who left everything to the Madonna of Loreto. The monstrance was made in 1699 by Baptist Kanischbauer and Matthias Stegner of Vienna. The monstrance, studded with 6222 diamonds, sends out its rays like a sun. It is almost three feet high and weighs more than 26 pounds (12 kg). Right next door to the Prague Loreto, the restaurant *U Lorety* is a pleasant example of a Prague garden restaurant.

Palais Czernin

If you leave the Loreto and walk on right up to the square, you are almost thunderstruck by the truly massive facade of the **Palais Czernin**, an incredible counterweight to the light building surrounding the Casa Santa, which, seen from this point, almost seems to cower. Twenty-nine half-pillars are here lined up one after the other and their height (more than two storeys) defines the palace front, over 490 feet (150 meters) long. In 1666, Humprecht Johann, Count of Czernin, bought the land, and work started on the palace straight away, under the direction of Francesco Caratti. In 1673, the Emperor Leopold I came to Prague and demanded to see the building about which there was so much talk in distant Vienna. It did indeed seem as if the Count, who had not received the imperial favor he expected, was building his own palatial residence out of pique. At any rate, the emperor was displeased when the Count announced that really it was nothing but a big barn

Window gallery in Novy Svet.

and of course he wasn't going to leave the present wooden doors in, but was going to replace them with bronze. "For a barn, those wooden doors are quite good enough", the emperor retorted.

The Czernin were an old Bohemian family and their members had excelled time after time in the service of the Bohemian crown. The house in Prague was to become a "Monumentum Czernin", but fate was no longer kind to the building. Building went on for several generations, until at length financial collapse put a stop to the project. It was partially destroyed during the sieges and wars, and in 1779, the heavily damaged building was to be sold, only no buyer came forward. During the Napoleonic Wars, the building was a military hospital, in 1851 the state bought parts of it and turned it into a barracks. In 1929, the young Czechoslovak republic had the palace renovated and made into the Foreign Ministry.

In the arcades opposite the Palais Czernin more mundane things are to be found. If you like dark beer, here in The Little Black Ox (*U Cerneho vola*), you will find the good and strong Velkopopovicky kozel 12 degrees.

Novy Svet

Below the gardens of the Palais Czernin runs a alley which belongs to the old settlement in front of the castle. In the middle of this former poor quarter is the **New World**, *Novy Svet*, which now draws many an artist and intellectual. The houses all have names, many have a house sign with the adjective "golden". This is something typical of Prague, and so the houses are called "The Golden Leg", "The Golden Star", "The Golden Pear". In **The Golden Pear**, *u zlaté hrusky*, is a romantic wine bar.

Veteran in Novy Svet.

THE ROYAL WAY

The last coronation in Prague took place on Sep. 7, 1836. Ferdinand V, the Austrian Emperor, became the last crowned King of Bohemia. Ferdinand V abdicated in 1848 in favor of his nephew Franz Joseph, who, however, did not allow himself to be crowned. Emperor Charles, who reigned from 1916-18, had himself crowned with full ceremony as King of Hungary—and in the middle of World War I too—but not in Prague.

People in Prague took umbrage at this behavior by the Habsburgs residents in Vienna; after all, ever since the imperial seat had been moved from Prague to Vienna, all of them had previously come to this ceremonial and important state occasion. And all of them had followed the route that is rightly called the Royal Way: from the **Powder Tower** to the **Celetná**, the **Karlova**, the **Charles Bridge**, the **Nerudova**, across the two big squares, the **Old Town** and the **Malá Strana** squares—even today, this route is designated the **Royal Way** by those who are officially responsible for the monuments of Prague.

Only with difficulty can one imagine the carriages with their four or six pairs of horses, the riders and runners proceeding through the streets. The buildings were for a long time subject to damage from the vibration of modern traffic, until they were turned into pedestrian precincts. Now the streets have quite a different appearance from how the Royal Way must have looked. Today the streets are dominated by shops and restaurants, by tourists and Prague families out for a stroll. What did it look like in earlier years? What did the people of Prague see when, eager for a spectacle, they lined the narrow streets of the Royal Way?

Preceding pages: facade of the Palais Czernin. Left, Maria Theresia. Right, "Charles IV" at the Mělník Festival.

Today you can follow one of these processions step by step and discover exactly where each salutation took place, how long the procession stopped, where the cheering crowds gathered. "...on both sides many thousand people of both sexes, all filled with joy, who continually cried 'Long live Maria Theresia, our most gracious Queen!' to give utterance to their rejoicing." On Apr. 29, 1743, on account of the "sudden inclement and windy weather", the queen had to travel in a closed carriage, drawn by six dark brown Neapolitan horses, strong and handsome creatures. To the left of Maria Theresia sat her husband Francis of Lorraine.

In those days the Horse Gate at the end of the Horse Market (now Wenceslas Square) was still standing. A great tent had been put up in front of

it, in which the queen had withdrawn for a short while. In the meantime the procession formed up, all 22 groups of it. All were on horseback, in new uniforms, the ladies in splendid coaches, accompanied by musical bands. After crossing the New Town in a wide curve, via Charles Square and Na příkopě, the procession came to the Powder Tower and the Old Town. From here on it went through the Celetná, passing what was once the most exclusive hotel in the Old Town of Prague, the "Golden Angel", and the Teyn Church with the courtyard behind it, then a hostel for traveling merchants passing through.

When the royal procession reached this focal point, they met in front of the Teyn Church, the deputation of all four faculties of Alma Mater Pragensis, and each nobleman or woman was greeted with a fine, well-turned speech—in Latin, of course.

Once the ceremonies of the Old Town Square were over, the procession passed the magnificent houses, several storeys high, that lined the route to the Small Square, and then entered the narrow Jesuit Alley (Charles Alley today), which winds around a number of bends till it meets the Crusader Knights Square. The last third of the way passed the Clementinum and St. Savior's Church. From the Old Town Bridge Tower, the triumphal arch spanning this route, the figure of its builder Charles IV looks down on the procession. Charles IV wears the imperial crown and the shield at his side bears the imperial eagle. At his side is the figure of his son Wenceslas wearing the crown of his saintly patron and bearing the imperial regalia in his hand. Between them, standing on a kind of model of the bridge, is St. Vitus, patron saint of the cathedral. The heraldic shields, lined up as if in greeting,

Family tree of Charles IV in Karlstein Castle.

display the badges of countries that were ruled by the king of Bohemia.

The **Charles Bridge** was not always the avenue of saints' statues who seem to be offering sound moral advice to those passing by. The coronation procession made its way over the Charles Bridge to the Malá Strana, this part of Prague that is so different from the narrow streets of the Old Town, and then entered the **Bridge Alley** between the Malá Strana Bridge Towers.

The Alley is short and quite broad, and above, high above the proceedings, rises the dome of St. Nicholas' Church. The procession passed along the south side of St. Nicholas' Church, the symbol of the Malá Strana and of the Prague Baroque style. Here the last part of the Royal Way began, the steep street that is now the **Nerudova**, which demanded all the attention of coach drivers and riders. It was particularly important to manage the great curve

The last curve before the Prague Castle.

into Hradčany Square smoothly and to keep the horses moving at an even pace.

The procession passed many houses with their delightful house names; the "Three Violins" or the "Two Suns", their signs above their gates visible for quite a distance. And yet two palaces have squeezed even into this stretch of road, that has been built on for over a thousand years. The first was commissioned in 1715 by the Morzin family from the famous architect Santini, the second built by Matthias Braun for the Kolowrat family.

From here, it took the royal procession only a few more steps to reach Hradčany. This is where the heir was presented to the people, a ceremony that partook of something of the nature of an election. The people were presented with the candidate to the throne, although the candidate's succession had long been established by the laws of inheritance.

PUBS AND COFFEE HOUSES

The "coffee house", that great Prague institution from the years before and between the two world wars, no longer exists. It used to be a place which had the latest newspapers, where the waiters were more like friends and confidantes, which was patronized by the "great" ladies of contemporary society and where the ambience was created by the flair of artists and journalists. There were big, ostentatious coffee houses in the city center—here the waiters, according to the writer Jaroslav Seifert, at any rate, went twice a day to the barber's to let themselves be shaved. Every coffee house had its "own" clientele—actors in the **Slavia**, Kafka, Kisch and the "Arconaut" circle in the **Café Arco**. Here couples met and held hands—over there was the meeting place of the demimonde. The coffee—and this still hasn't changed even today—was well-known to be awful, and you paid your two crowns for it more or less as an entrance fee. In winter you came to get warm and to save on heating costs. In summer you came "for the thick tobacco smoke"—according to Seifert. The number of hours that many a customer wasted in these coffee houses while life passed them by—it doesn't bear thinking about.

Even if the glories of times long past are no more to be found here, and the "Bohemian" lifestyle seems to be over—the coffee houses of Prague are still worth a visit. Not only because of the interiors, which seem so nostalgic nowadays, with art nouveau decor such as that in the **Europa** or the **People's House**.

The last of the large and important coffee houses in Prague was and remains the **Slavia**. Nowhere else, except in the Slavia, will you find such a cross-section of coffee house customers. They may not come as regularly as they used to, but the actors, artists, and singers still come to the Slavia from the opera and theater opposite. And it can still happen that a slightly greying prima donna will hold court among her friends. The eternal, timeless artistic stereotype—smoking heavily, wearing a beret, clutching a manuscript—can also be seen in the Slavia. With a bit of luck, he'll let you pay for the wine and in return tell you some of the inner secrets of literary life in Prague. But in the Slavia you can also find the old ladies, factory owners' widows from the First, the "Golden" Republic. However, they have become a rarity in the Slavia, and meeting such a lady is quite an occasion.

But young people also come in, from the nearby Conservatory, to drink coffee in-between two lectures or to meet their friends. And in the evening the café is a meeting place for

Preceding pages and left: in the Café Slavia. Right, a favorite place in summer—U Fleku.

"yuppies"—if there could be said to be such a thing in Prague—well-dressed, with hairstyles á la David Bowie and Benetton clothes from the Tuzex shop. Next to them, in customary black, sit those from the opposite end of the fashion spectrum: punks with colorful, wild streaks in their hair and heavy army boots.

The waiters may no longer go to be shaved twice a day, but they haven't lost any of their pride in their profession. They don't react to pressure or impatience from their customers, and many a tourist has to suffer a long wait.

You won't find the coffee house atmosphere in the days of the tourist season, when seemingly endless streams of tourists descend on Wenceslas Square. If you want to see where the young men meet, or listen to a well-known, very elderly transvestite telling stories of better days, you should go to the **Europa**—but only in winter.

At any rate, the coffee houses are and hopefully will continue for a long time to be established institutions in Prague.

Pubs and Beer

Many people in Prague can tell you tales of their grandfathers, who drank 20 or 30 tankards of beer a day, played cards in the pub and never so much as pay a visit to the gents' room. These tales are, of course, exaggerations. What remains is the love of beer. Yes, Cedok does offer "beer parties" in Prague—organized merry-making in the Interhotel. This has nothing to do with the life in the pubs of Prague. A simple restaurant or pub doesn't have to be clean, as long as it serves the right beer, Smichover or Pilsener, each to his own. And it has to be fresh, the foam mustn't collapse straight away, and the draught taps have to be cleaned regularly. It is said that you get better

Pub brawl, **painted by Josef Lada.**

beer in the **Koruna** in Wenceslas Square, which is nothing but a snack bar full of vending machines, than in many a "traditional" pub or restaurant. Where do they have especially good beer on tap? One of the best places is still the **Little Black Ox** up in Loreto. Here you'll still find many Prague "characters", with broad middles and loud voices, who drink their beer with such speed that a stranger can hardly keep pace. Or just try one of the pubs around the corner, most of which are packed full. What about a beer garden in *Riegrovy sady*, where a band plays on the weekends in the summertime and people come in from the surrounding streets to enjoy a little dancing? Of course, you mustn't miss the **Chalice** with its mementos of Schwejk or the **U Fleku** with its beer garden. Another good place is the **Golden Tiger**, in which you may, if you're lucky, find one of the most famous Czech writers.

Window on the Café Slavia.

Wine Bars

The Malá Strana wine bars were once simple places, where one met, got something to eat—just a small snack—and whiled the night away over good wine. But here also one or two changes have taken place. Wine bars today are often very formal and expensive restaurants, with a plethora of white napkins and starched tablecloths. Only a few have profited by being changed into pretentious tourist traps.

However, if you don't want to eat in grand style—in the **Lobkowic Wine Bar**, for instance—but just drink a small glass of wine, there are still some places to go; the **U Golema** and the **U Rudolfu** in the Maiselova, for instance, or the "Golden Frog", **U zlaté konvice**, in the Melantrichova. The most famous and popular wine bar in Malá Strana is still the "Maecenas", **U mecenáse**, in the Malá Strana Square.

NOT JUST FRANZ KAFKA AND SCHWEJK

The "Golden City" on the Vltava has always been a golden fertile ground for poetic language. In the early 20th century German and Czech had equal status in Prague, and each produced a world-famous novel: **Franz Kafka's** *Josef K.* and **Jaroslav Hasek's** *The Good Soldier Schwejk*.

In those times, marked by the turn of the century and World War I, by literary anarchism and avant-garde, the Prague literary scene was formed, which, with an added dash of "Bohemian" lifestyle, was to culminate in the 20s and 30s and even grow to rival Paris.

"Uncontrollable, in a certain sense"—this quote from Kafka about an unexpected love may perhaps point to a strange phenomenon: authors writing in Czech rarely seem to have discovered anything sinister or secretive when they wrote lovingly of the city on the Vltava; authors writing in German, however, had a tendency to imagine dark secrets into their view of Prague and to prefer tragic subjects. **Nora Fryd**, who was born the German Jew Norbert Fried and became part of the Czech literary scene under the Czech version of his name after a "wretched and blessed" period of his life spent in Theresienstadt, Auschwitz and Dachau, was asked to comment on this phenomenon. His reply: "The German-speaking authors, Jews or non-Jews, often concentrated on the dark events in the varied history of Prague and imagined tormented personalities. Jews or non-Jews, they turned to the Golem and other legends about or based on Rabbi Löw, or to the secrets surrounding Rudolf II's cabinet of curiosities. The Jews of the 20s probably felt hints of their terrible fate in 1938, the German-speaking non-Jews perhaps had an inkling of their ill fortune in 1945. Those non-Jews and Jews who wrote in Czech, however, preferred to see in Prague only its golden attributes, Prague as Maticka (Mama)." Another quote has come

down from Kafka: "Mama has claws!"

There were several cafés and pubs in Prague in which, occasionally and temporarily, the Czech, Prague German, German-Jewish, Czech-Jewish or Sudeten German authors met. However, Max Brod adds a caveat about the most famous of these places: "The tales told about the Café Arco are untrue or at least highly exaggerated. Not until Werfel and his clique made the new Café Arco into their "local" did it amount to anything." Well, the coffee house that

Werfel, Kafka and Brod knew and frequent no longer exists.

Rather than searching for the places where the liquid intake crystalized the literary genius of Prague, it is probably more productive to let the city, indeed the combination of the landscape and the city, work on you as a whole. This is how the poetry school of the **Prague Circle** came into being, which, as Max Brod claimed, had no teachers and no curriculum. The *spiritus rector* added: "Unless, you might claim, Prague itself, the city, its people, its history, its beautiful surroundings, the forests and

Left, portrait of the "good soldier" Schwejk. Above, Franz Kafka and Felice Bauer.

181

villages that we walked through on enthusiastic hikes, was both our teacher and our curriculum." No doubt this education in openness and tolerance made it possible for Max Brod to support the Czech Hasek as much as his friend Kafka. It's well-known that Kafka would never have attained his position as the classic author of modern literature if Brod, against the express wishes of the author, had not rescued and sorted Kafka's manuscripts, thus saving this important work and making sure that it spread from Prague to the whole world. He had a similar success with Hasek—i.e. he

idiom of Prague, the German dialect current in the Malá Strana, and so converted the play into German with great originality. The German translation was Hasek's first step on the way to international literary recognition. All over the world people were delighted by the cunning dog handler, who found his way out of the "historic situation" of World War I through "nonsensical behavior and a mask of clownishness" (Radko Pytlík).

It must be said that the Czechs are no longer entirely happy about the success of this book. After all, it led to the internationally current prejudice that all

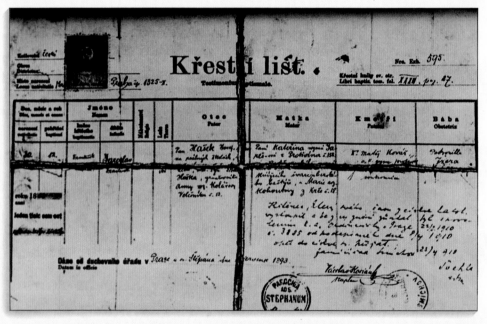

made him into a classic author of modern Czech literature by internationally promoting Hasek's novel S*chwejk*.

It seems incredible today, but then, in the 20s, the Czech "good soldier" of World War I did not find favor with the Czech intelligentsia, who considered him to be immoral, improper and damaging to the Czech national image. Even so, Brod sent a dramatized version to Berlin to **Erwin Piscator**, under whom **Bert Brecht** was working at the time. The successful translation into German by **Grete Reiner** helped the play to succeed. She used a native

Czechs are like Schwejk, or, at the very least, there is a bit of Schwejk in every Czech.

Jaroslav Hasek was a Bohemian, in both sense of the word, of the first order. Within the space of an hour he could switch from complete concentration to apathy, he could work steadily towards a goal, yet within his lifespan of only forty years he moved the goalposts several times at a moment's notice. He lived through it all, from anarchist gestures via plans for an imaginary Romanov on the Bohemian throne to work for the communist party, from the foundation of a "Party for gradual progress within the framework of the law" to alcoholic apathy. He lived through it all and included it all in his humorous and satirical masterpiece. Hasek was the literary creator of forgotten and of timely characters. His creative source was the pub, and he included many a quote from the bar. He immortalized the pub *U Kalicha* (The Chalice) together with the good soldier Schwejk, and thus left Prague with a meeting place for natives and visitors who want to combine alcoholic spirits with the spirit of literature.

However, modern literature in Prague meant not only Kafka and Hasek, Brod and

arcades and in the archways, on the bridges and in streets such as those between the Waldstein and the Malá Strana squares, in front of St. Wenceslas' statue and the Hotel Europa (then the Grand Hotel Schroubek).

The Czechs in Praha had in the 19th century already known their **Jan Neruda** and their **Bozena Nemcová**. The *Tales of the Malá Strana* and the *Grandmother* had found their way, in translation, into German bookcases as well. The modern age of Czech literature developed with **Josef Capek, Jaroslav Durich, Frantisek Halas, Vladimir Holan, Josef Hora, Josef Lada,**

Capek, Kisch and Nezval. Whole galaxies of literary stars, writing both German and Czech, enlivened the city on the Vltava with their texts and their charisma, with their books and with their personalities. There may not have been a Montmartre as in Paris or a Schwabing as in Munich, yet the writers filled the capital of the first Czechoslovak republic with an ambience that can still be felt with surprising immediacy, under the

Left, Jaroslav Hasek's birth certificate in the U Kalicha. Above, Karel Capek is buried in the Slavín in the Vyšehrad.

Frantisek Langer, **Marie Majerová**, **Ivan Olbracht**, **Jaroslav Seifert**.

There's no doubt that there was little contact between the authors who wrote in German and those who wrote in Czech during the years between the wars. The unhappy nationalism of the 19th century continued unabated in post-1918 Czechoslovakia. Of course individual authors built literary and political bridges. Max Brod could still speak friendly words in 1966, although he was only referring to the last few years of the CSSR. He praised the exceptions: "The walls of isolation were breached. There were, after all, many points of contact. There was a "social club" in one of the palais on Na příkopě which was open to both languages and subsidized by the government. Also, German speakers went to the Czech theater and concerts and vice versa. It was natural for some German papers (not all) to cover fully all the events in Czech cultural life (theater, music, art, literature)—and vice versa too." Brod believed that before 1938-39 he could see on the horizon the possibility of the two cultures working together in a European context, in particular when **Max Reinhardt** produced **Frantisek** Writers' Association was to be found in no.11, and in no. 9 the main publisher for literature written in Czech, the *Cechoslovensky spisovatel*, together with **Odeon**, the main publisher foreign literature, and **Albatros**, the main publisher of books for children and young people. Where Albatros is now, used to be the site of the Café Union, in which writers and artists of almost every field met—but it no longer exists today. You can find Albatros quite easily by following the neon signs in the Národni trída: NEJLEPSI DETEM (the best for children). This refers to the self-imposed

Langer's *Periphery* in Berlin, but in the end he had to admit: "Unfortunately the possibilities were only hinted at. The point where the parallel lines would meet was never reached."

At the end of the 40s, when the CSSR became a socialist country, literary life was monopolized by the demands for social realism in literature. Publication was concentrated in relatively few publishing houses and journals, and the influential Czechoslovak Writers' Association was formed. Now literary activity in the city was concentrated in the Národní trída, where the

duty of the authors, illustrators and editors to provide children and young people with nothing but the best books.

Josef Lada, whose illustration of Schwejk is recognized throughout the world, and **Jirí Trnka**, whose pastel drawings with their childlike appeal became a successful export, used to be familiar faces at Albatros. Now they have been replaced by **Ota Hofman**, whose *Pan Tau* has charmed children everywhere, and **Otakar Chaloupka**, whose ideas on literature for children and young people have had considerable influence.

Visitors to Prague who are looking for personal traces of and contact with literary figures are very likely to meet one or other of the key creative figures in the Czech literary scene somewhere in the Národni trída. Some of them can be immediately recognized as such just by their appearance. The well-dressed grey hair stands up slightly from the scalp, the clothing is both neat and casual, the expression soft and the posture strangely upright and bowed at the same time. They go in and out through the narrow doors of nos. 9 and 11, and hurry to their appointments—in the editorial offices of the publishing

house or in the restaurant of the "club" of the Writers' Association, which is unfortunately only open to members. In the years before the war, this "Club" was the Café National (*Národní kavárna*). In the years just after the war the "Club" still had guests such as Nezval, Halas and of course Jaroslav Seifert, the Nobel Prize winner of 1984. You could add name after name from the "Prague Spring" of 1968. Some of them are no longer

Left, in the Baroque hall of the library in the Clementinum. Above, Egon Erwin Kisch, the "rushing reporter".

entitled to be members of the "Club", and some live or publish abroad.

However, most have stayed, in the home country of their language, in their city of Prague. Sometimes they drink coffee in the Café Slavia, a few yards on towards the Vltava, next to the National Theater, where Jaroslav Seifert found inspiration for the "Café Slavia" poems:

Through the secret door from the Vltava quay,
which was of such transparent glass
that it was almost invisible,
and whose hinges
were smeared with oil of roses,
Guillaume Apollinaire would sometimes come.
His head was still bandaged,
from the war.
He sat down with us
and read beautiful, brutal poems,
which Karel Teige translated on the spot.
In honor of the poet
we drank absinthe.
It was greener
than any other green,
and if we looked from our table
through the window,
the Seine flowed past the quayside.
Ah yes, the Seine!
And not far away, on broad-spread legs,
the Eiffel Tower stood.
Once Nezval came, wearing a stiff hat.
We didn't know it then
and he didn't know either,
that Apollinaire was wearing the same one
when once upon a time he fell in love
with the beautiful Louise de Coligny-Chatillon,
whom he called Lou.

The star among the writers of today, however, is **Bohumil Hrabal**, the literary creator of the "Bafler", whose *Uncle Pepin* has brought another Schwejk to life. The "star" Hrabal holds court, democratically, in the pub "Zlaty tygr" (The Golden Tiger), not far from the Národní tr., and anyone is welcome to come and chat with him.

PRAGUE—CITY OF MUSIC

Bohemia has a rich musical past, and it has brought Prague the reputation of a "musical city". However, the city—the "musical heart of Europe", as David Oistrakh once said—doesn't just owe its reputation to the "big names", which after all grace many other cities. Rather, the important position that is accorded to Prague by statements such as Oistrakh's is based on a long tradition of musical culture.

Bohemian Musicians

The great flowering of Czech music, also known as "Bohemian classicism", took place in the 18th century. The saying "All Bohemians are musicians" dates from this time. The proverbial musicality of this nation is probably due to the fact that in Bohemia, support for musical education was widespread. Documents dating from this time show that most cantors (school-teachers) had a musical education and saw to it that nearly every pupil could play an instrument or at least sing.

The *General Musical Journal* of 1800 says: "A great number of these cantors were truly skilled and talented musicians...The vast number of skilled Bohemian musicians, of which one can find no better proof than by reading the lists of players in the court orchestras of Europe, is explained by the fact that the best of the nobility insist that all their servants—from estate manager down to stable boy— should enjoy music and be able to play an instrument properly."

Musical talent and education, in those early days, was an excellent means of obtaining material comforts and advantages. A position as a servant freed a peasant from serfdom and from military service, and those who proved themselves to be good musicians had the hope of being released from the service eventually.

When the English music writer Charles

Burnley visited Bohemia in 1772 he was so surprised by the level of musical knowledge and skill in the country that he named it the "conservatory of Europe". The fruitful musical climate produced not only many folk musicians but also a surplus of trained performers, who had difficulty in earning their living in their home country.

The unstable political situation in the Bohemia of the 18th century and religious persecution forced many people to emigrate. Countless musicians also left the country, and found work all over Europe on account of their skills. Everywhere these emigrants went they commanded respect, influenced the new instrumental style of Classicism, and left definite traces in the structure of its melodies. On the other hand, Bohemian music became exposed to foreign influences, which it in turn incorporated.

Mozart in Prague

The visits of Wolfgang Amadeus Mozart, who found many friends in Prague, should also be seen in this context. Mozart tried to build up a secure career for himself in Vienna, but the Viennese public and the imperial court mainly reacted with incomprehension and indifference. At this time he received news from Prague about the reception of his opera *The Marriage of Figaro*. An invitation quickly followed, and which he took up in early 1787.

In Prague he witnessed Figaro-fever, which had gripped the whole city. Apart from this, the visit brought Mozart a contract for an opera. It was to be *Don Giovanni*. It was commissioned by the impresario of what was then the Nostitz Theater (now the Tyl Theater), which, in contrast to all the other theaters in Central Europe, was not tied to a court, but was a relatively independent institution. The fact that in Prague opera had been available to the general public for a long time explains the interest of the broad mass of people. The premiere in the autumn of the same year was a great success.

Music and Middle Classes

From the early 19th century on, the aristocracy of Prague gradually lost their position as the most important patrons of the arts. The rising middle classes claimed their share in the process of shaping cultural life. The center of activity moved from aristocratic salons to public concert halls, and a new era dawned. It was formed by two institutions which left a definitive mark. One was the Society of Artists, founded in 1803 and modeled on its predecessor in Vienna, the other was the Prague

Vltava island (*Slovansky ostrov*) became a venue for Hector Berlioz, Richard Wagner and Franz Liszt. Liszt also played in the Platyz *(Uhelny trh 11)*. The scene was set by the mighty flood of largely German music, and Czech music faded into the background.

Smetana and Dvorák

The awakening of Czech national consciousness during the politically and economically troubled times of the early 19th century saw the first generation of Czech artists faced with the task of creating

Conservatory, which opened in 1811, was the first in Central Europe and set the standards for the rest. The city, which was still under the strong influence of the Mozart cult, was exposed to new influences. Carl Maria von Weber, who was director of the Nostitz Theater from 1813-16, acquainted Prague with Beethoven's *Fidelio* and the first Romantic operas. In the same house Niccolo Paganini celebrated great successes. Concerts also took place in the Konvikt, a complex in the Bartolomé jska ul. which is currently being restored. Beethoven appeared here. Later, a concert hall on the

their own culture, which did not establish itself until the second half of the century.

The name of Bedrich Smetana is inextricably bound up with Prague, and in his work Czech music reached its first peak. Smetana came to Prague to study music. During the Czech nationalist rebellion in 1848, in which he personally took part, his patriotic feelings awoke. His wish was to unite the highest artistic expectations with the demands of an independent national culture. However, the way to this goal, which he was to achieve most of all in his operas, was long and difficult. Apart from a five year stay in Goteborg, Smetana took part wholeheartedly in the musical life of Prague, but at first he tried in vain to establish himself as a conductor and composer. Not until the success of his opera *The Bartered Bride* did he achieve the desirable position of conductor to the Czech Opera and widespread recognition, which, however, did not remain uncontested. After the loss of his hearing Smetana had to give up his career as a practising musician, but continued to compose and created some notable works. When the National Theater was opened, a ceremony that symbolized the

Dvorák, a full-blooded musician, whose never-failing wealth of ideas was generally envied and who was a true child of the musical traditions of Bohemia. Dvorák had great influence on musical life in Prague, as a conductor, as professor of composition at the conservatory and later as its director.

Modern Music

However, the strong flow of national culture did not have a detrimental effect on Prague's open-minded attitude to modern European music. Gustav Mahler, in 1885

peak of national aspirations, Bedrich Smetana received the highest honor when his opera *Libussa* was performed at the official opening.

While arguments raged in Prague about Smetana, a young Czech composer began to draw attention to himself. Soon he became famous outside the borders of his country, and through him Czech music was to achieve international renown. This was Antonín

Left, bust of Antonín Dvorák. Above, the former Nostitz Theater.

conductor of the New German Theater's orchestra (today it's the Smetana Theater), had the first performance of his 7th Symphony take place in Prague. The same theater was directed from 1911-27 by Alexander von Zemlinsky, who had close contacts with centers of music in Vienna and Berlin and acted as a go-between. By this means Alban Berg and Arnold Schoenberg, among others, had the opportunity to get to know the Prague music scene. One result of these busy cultural exchanges was the premiere of Schoenberg's *Expectation*, which took place in Prague.

CONCERTS IN PRAGUE

If you want to hear classical music in Prague, you won't be disappointed. A varied program is assured by several symphony orchestras, a number of chamber music ensembles, two opera houses and many soloists, together with visiting foreign musicians. Up-to-date information is given by the Prague Information Service (Na příkopě 20) and posters put up all over the city. The traditional repertoire is dominant in the programs of the (always well-filled) concert halls, and Czech composers, both old and new, feature prominently. Opera performances are somewhat overshadowed by instrumental music, which everywhere lives up to the highest expectations. For those who like chamber music, the various string quartets, among other groups, can be strongly recommended. They are remarkable for their musical excellence. Regular series of concerts of course also feature those musicians based in Prague who have received international recognition. Among them, without any doubt, is the Czech Philharmonic.

Musical events always lead the visitor to interesting places in Prague. In this way he is offered an excellent opportunity of experiencing many of the "sights" in a quite different way, of seeing the interiors and of spending time in places which are definitely worth seeing. Apart from the great concert halls and the opera houses, these are usually churches, palaces and palace gardens. For example, the effect of the bold architecture of St. Vitus' Cathedral is enhanced by the addition of a large orchestra and choir, and the polyphonic music of medieval times re-awakens the Romanesque St. George's Basilica to new life. The bare, impersonal interior of the Bethlehem Chapel suddenly loses its museum-like character when it becomes a concert hall, and the Baroque splendor of St. Jacob's Church in the Old Town only really unfolds during the organ concerts, which take place regularly every Tuesday afternoon. Historic rooms that are normally inaccessible are opened for musical events, for instance those in the newly restored Palais Martinic in Hradčany Square, or the Great Hall in the Palais Waldstein, the mirrored chapel of the Clementinum and others. Music draws people to the Slovansky Ostrov island, to the Riding School in the castle, to the St. Agnes Convent complex, or to the Hvezda summer palace (Letohrádek Hvezda—on the western edge of town). In summer open-air concerts are added to the list, in the gardens of the Palais Waldstein, in the Maltese Gardens of the Museum of Musical Instruments in the Malá Strana, or in the palace gardens below the castle. In the Baroque summer palace of the Villa Amerika, the Dvorák Museum, you can listen to music surrounded by mementos of the composer, as you can in the Villa Bertramka and its garden, where Mozart completed his operas *Don Giovanni* and *La Clemenza de Tito*. Mozart's venue in Prague, the old Nostitz Theater (now the Tyl Theater) is shortly to reopen after restoration work. In the "Music Theater—Lyra Pragensis" (Opletova ul. 5), performances take place which take themes from a great variety of musical experiences and present them to the public in the form of film or sound recording.

Once a year the posters in Prague are dominated by a white f on a blue background. The symbol, which looks like the sound aperture of a violin, announces the "Prague Spring". The musical life of Prague culminates in this festival, which has a tradition going back over 40 years and a fixed place in the international calendar of festivals. May 12 marks the anniversary of the death of Bedrich Smetana (1824—1884), the founder of the Czech national music movement of the 19th cenury. This cultural event is regularly opened with a performance of the composer's *Ma Vlast* (My Country), a cycle of symphonic poems, composed between 1874 and 1879. The best known of the pieces is the second poem, *Vltava*, which describes the course of the Vltava river.

TENNIS IN CZECHOSLOVAKIA

Following its successes over the last twenty years, the reputation of the Czechoslovak tennis style is recognized throughout the world. But only experts and true tennis fans know that the roots of Czechoslovak tennis lie much deeper and date back to the 19th century; they share the same age and the same traditions as, for example, those in Britain.

The oldest club is the CLTK, founded in 1893, which was a member of the UK Tennis Association from 1894-1906. The most famous players of earlier times were K. Kozeluh, professional world champion in 1929, 1932 and 1937, and J. Drobny, who won the Wimbledon championship in 1954 and reached the final in 1952 and 1949. V. Suková, the mother of Helena Suková, also reached the ladies' final in Wimbledon, and Jan Kodes won the men's final. It is impossible to imagine modern tennis without names such as Martina Navrátilová, Ivan Lendl or Miloslav Mecír.

After Jan Kodes' Wimbledon victory in the men's final in 1973, tennis became very popular in Czechoslovakia. Young people were encouraged to play. Matches were organized under the heading "The Search for New Kodes' and Sukovás". More than 15,000 young people aged between 9-15 took part.

Today the number of players organized in clubs has reached 60,000, and perhaps another 60,000 play for fun. There are 810 clubs with 3,500 places in the CSSR. Special training methods, long-term tuition of young players and the high standard of the Czechoslovak coaches have given the CSSR an important position in today's tennis world. Czechoslovakia has its own system of long-term tuition for young people, who start to play as young as seven years old.

To explain the success of the Czechoslovak style of tennis, a reminder of a few facts about the most notable players may be necessary. **Martina Navrátilová** was born on Oct. 18, 1956 in Prague and grew up in Revnice, a village not far from Prague. Her first coach was her stepfather. She started her career as Junior Champion in 1972, and in 1973 she was in the junior final at Wimbledon. In 1975 she left Czechoslovakia and became an American citizen. Martina Navrátilová was the third woman to win the "Grand Slam" in Wimbledon eight times. In 1984, Navrátilová won 74 matches one after the other and was defeated in the 75th by another Czechoslovak player, Helena Suková.

Ivan Lendl is proof that Prague is not the only source of good players. He was born on Mar. 7, 1960 in Ostrava in northern Moravia and in the 70s he was the most famous player in the CSSR. He was awarded the title "World Junior Tennis Champion" in 1978. Lendl's talent for tennis is the work of his parents. His mother was a successful tennis player, who became Czechoslovak champion in the doubles in 1964 and 1969. Jimmy Connors once remarked that Lendl looked "like a half-boiled chicken". Lendl's answer was plain: he beat Connors at Flushing Meadow and had a half-cooked chicken delivered to Connors' room. Lendl has not yet achieved his great dream, a Wimbledon victory.

Jan Kodes was born on Mar. 1, 1946 in Prague. In 1964 he was Junior Champion of Czechoslovakia, in 1973 he won in Wimbledon. In 1980 he and the team of the CSSR won the Davis Cup. For 11 years he was the no. 1 of Czechoslovak tennis. Today Kodes, who has done a great deal for tennis, is director of the new tennis stadium in the Stvanice district of Prague.

PUNKS IN PRAGUE

Yes, they do exist—punks in Prague. You may not find quite such a conspicuous youth scene as in many cities of Western Europe, but you do find traces of most youth cults. But you will notice that there is no individual scene for Prague, nothing typical of the city. What there is, it is mostly copied from Western models.

Music is one of the great refuges of the younger generation, and not only in Prague, but in the whole of Czechoslovakia. Most young people are no longer interested in political protest or ideological discussion. Most of them have grown up in the years following the Prague Spring. This period—which brought about a mood of resignation among most of their elders—also influenced young people. Skepticism and withdrawal, the retreat into their own private world is the main concern of young people in Prague.

Obviously, not all young people have withdrawn from the mainstream to create their own secure little world. Young people are as divided as the rest of society. The time of the broad unified youth movements is past, in the West and in the East. In the CSSR and in Prague, too, the scale reaches from disco to punk, from New Romantics to Heavy Metal.

Discos in Prague

Prague does have a disco and nightlife scene—that soon becomes obvious to anyone visiting Wenceslas Square. The latest pop videos from abroad are shown in the video discos, and, just like in the Golden West, a bouncer is necessary to sort out those who can come in from those who can't. However, you shouldn't expect too much from a video disco. Often there are only one or two TV sets fixed above the dance floor. "No jeans or sneakers" say the signs on the door of the **Discotheque Zlatá Husa**, entrance fee 30 Kčs. You can dance to rock music till the small hours in the **Jalta Club**, in the **Hotel Jalta** (also in Wenceslas Square)—if you pay the 50 Kčs entrance fee and remember that it is a very expensive place (1st price bracket plus 30 percent). However, the place is mainly considered to be a meeting point for Western tourists and black market currency dealers.

The **Video Disco Alfa** on the opposite side of Wenceslas Square is better for meeting people, or you can go to the **Tatran**, which has a plain glass floor lit with colored lights. Here you still have to pay your 30 Kčs at the door, but the atmosphere is much more relaxed and friendly than in the Zlatá Husa or the Jalta. The discos in the **Boatels** (*Nábr. L. Svobody*) are also very popular with young people in Prague. The **Admiral** and the **Albatros**, for instance, are both in price bracket B, which makes them more affordable for both East and West. If you really want to get away from tourists for once, try the **U Holubu**, Praha 5, S.M. Kirova. For 10 Kčs you can really let your hair down and dance till two in the morning. If your taste runs to punk or New Wave, take yourself off to the **Na Chmelnici**, Praha 3, 10 Konenova. Some members of this scene meet at the monument to St. Wenceslas. And with a bit of luck you might find an improvised memorial in Old Town Square with the inscription "Lennon" rather than "Lenin". Of course things of that sort don't remain standing for long.

In summer Charles Bridge is another favorite meeting place. Young people meet here to play the guitar together and have a short jam session. There's no rigid division in Prague between the various "scenes". There are no pubs exclusively for teenagers or for punks.

Right, you can hear Heavy Metal sounds in Prague too.

You meet your friends and go to any old pub in the lowest skupina (price bracket). If this is redecorated or promoted to a restaurant, you go somewhere else. In the outlying districts of Prague there are smaller discos or places where young people meet, which advertise by putting up posters on the fences around building sites.

Heavy Metal fans who want to listen to the Czech Hard Rock groups Citron or Vitacit should go every Thursday to the **House of Culture**, Kulturní dum Barikádniku, Saratovská 1, Praha 10. Here they'll also find a Heavy Metal disco in the Hall of Culture. Most of the concerts of this sort take place here. Beer and Becherovka flow in rivers, and sometimes the place is really buzzing.

Of course concerts don't always have to be loud and heavy. In Prague young rock groups such as Stromboli, ETC or Vyber are also very popular. However, there is no one place in which to hear these groups. Your best bet is to ask in the **Sluna** box offices in the Alfa Passage or in the Lucerna Passage in Wenceslas Square. Here, for 10-20 Kčs, you can get tickets for all sorts of events, rock, jazz, chamber orchestra or lieder recital. Occasionally small rock concerts take place in the **Lucerna Palace**. You can get a ticket to see three to five groups for 20 Kčs.

Folk and Jazz

Concerts on a larger scale take place in the **Palace of Culture**, *Palác Kultury* (Metro station Gottwaldova), or in the Park Kultury a oddechu Julia Fucíka, Praha 2 (Metro station Vltava). In summer free open air festivals lasting two days are held here. In April a rock festival lasting several days is held in the Palace of Culture. Folkrock and blues concerts are also very popular. Country, folk and traditional folk music

After a pop concert in the Prague House of Culture.

concerts are held in the **Sophia Hall** on the Vltava island of Slovansky ostrov.

The very best jazz can also be heard in Prague, with bit of luck. From time to time the Prague Jazz Days are held, and musicians of international repute are among those who attend. However, the local jazz scene also has plenty to offer.

When Miloslav Svobada invites his guests to the **Malostranská beseda** (the Malá Strana House of Culture), the occasion is always a huge success. This place, right in the Malá Strana square, is one of the best addresses for good jazz in the whole of Prague. Miloslav Svoboda, Martin Kratochvil and Jana Koubkova regularly appear here. There is some sort of event every two or three days in the Jazzclub. The second best address for jazz in Prague is bound to be the **Club Reduta** in the Národní trida. A somewhat lower quality is on offer in the cellar bar on the other side of the street, the **Metro Jazz Club**. Large-scale events such as the "Jazzparanto" with Jana Koubkova take place in the Palace of Culture. The **Jazzlokal Prazan**, in the Julia Fucika Park, puts on a variety of jazz events.

Who's Who

"New Romantics" and the Prague yuppies meet in the Grill Room of the **CKD Dum**, right by the exit from the Metro station Mustek in Wenceslas Square. It's de rigueur to be over-dressed in the best designer clothes, just like their counterparts in the West.

If you like art nouveau and prefer the company of young men, the Café Europa is the right place to go. This coffee house is open from six a.m. to midnight and isn't only a place for making this sort of contact. Just as in the other great coffee houses in Prague, the Café Slavia's clientele is very colorful and mixed.

Roller skating is summer training for ice hockey players.

AVANT-GARDE ARCHITECTURE IN PRAGUE

Curlicues to Art Nouveau

In the Europe of the turn of the century, modern styles of architecture were evolving. Prague was, along with Paris, Vienna and Berlin, one of the places where the artistic avant-garde were using revolutionary ideas and manifestos to shape architectural styles so that they reflected their ideals of the world. A progressive atmosphere, such as one can hardly imagine today, was created by close economic and cultural ties between the cities. Architects from Germany (Peter Behrens) and Vienna (Adolf Loos) designed important buildings for Prague. Le Corbusier, Walter Gropius and Hannes Mayer held influential lectures, and Czech architecture won itself a place in European artistic journals. In this environment an extremely creative, internationally influential and yet Bohemian-inspired architecture could develop to mark an important and fascinating chapter in the history of Prague. It is sometimes more exciting and challenging to discover the monuments of this overturning of the artistic establishment than to visit the monuments of long-vanished ages.

As in Berlin and Vienna, modern architecture in Prague developed out of the Historicism left over from the 19th century, which gives whole districts of Prague their characteristic appearance. The first liberation from this style, and the beginning of the modern period, was the advent of art nouveau with its natural style of ornament, instead of the over-used curlicues of Historicism. The most beautiful example is considered to be the **Hotel Europa** (1), which is built in 1900.

The man who did most to initiate the U-turn in architecture was the Viennese master architect Otto Wagner, whose student Jan Kotera was one of the first to introduce new trends into Prague. At first he was strongly influenced by Vienna, as can be seen in the **Peterka House** (2), built in 1900, with its tall slender windows and elegant art nouveau ornamentation; he later combined new ideas with Bohemian characteristics. He created a new architectural language, as in the **Urbánek House** (3), built in 1912, where ornament has retreated into the background in favor of the effect of the material, bricks and copper.

The Stenc House (4) by the Kotera pupil Otokar Novotny, with the finest of brick facades built with exquisite restraint, dates from the same time. Here, architecture is not understood as a matter of ornament, but as a play of proportions, light and shade.

The Wagner pupil Hubschmann marked the courageous transition to a plainer architecture more suited to the new times with his **apartment house by the Jewish Cemetery** (5), built in 1911. Further development of Kotera's ideas can be seen in the **office block** (6) built in 1924, in which Kotera consciously incorporates Baroque forms into his building.

Cubism and Rondocubism

The second chapter of modern architecture in Prague began with the exhibition of the first Cubist paintings (Picasso, Braque) in Paris. These changed not only the development of painting, but also the style of buildings in Prague. In the revolutionary atmosphere of the times, people saw Cubism, with its faceted and prism-like dissection and abstraction of surfaces, as a possible way of overcoming old conventions and projecting an image of the new age. **Josef Chodol** became a notable architect of the times with his buildings in the **Necklanova 2** and **34**. Another example from the inner city is the **street lamp with seat** (7) constructed in 1913 by Vladislav Hofmann, which is delightfully linked with the church of St. Mary of the Snows.

An early example of the period dates from the year 1912. It is the **House of the Black**

Mother of God (8), by Josef Gocar. This building, by the way, has a lovely café on the first floor, which is well worth a visit.

The Cubist influence came at a time when the country was in the process of dissolving its ties to the Habsburg monarchy and founding the Czechoslovak state. This was the political background to the efforts of leading architects, who were trying to create an independent national architecture using Cubist methods combined with forms of vernacular architecture such as arches, cylinders and shapes in high relief. This style is known to art historians as **Rondocubism**

and has had a strong influence on many buildings in Prague. Notable examples of this style are the **office block** (9) by Pavel Janak, dating from 1922, and its neighbor dating from 1923 (10), shortened to give a monumental perspective and by the same architect. Both of these buildings are situated opposite a house by Kotera which was built some ten years earlier.

The master architect, Novotny, also designed numerous buildings in this particular style, for example, the **hostel** (11) with its new-style, and very effective, pleasant color scheme.

Modern Architecture

During the course of the 20s, European architecture took another direction, and the white "classic" style of modern architecture prevailed over the nationalist rondocubist style. The third chapter of modern architecture in Prague began. A symbol of this new direction, and of the cosmopolitan character of Prague at that time, is the light, almost disembodied appearance of the **Manes House of Artists** (12) dating from 1928. This radiant, white, puristically plain building was confidently designed by Novotny to protrude into the Vltava.

An interesting succession of interiors dating from this third period is provided by the shopping arcade **Black Rose** (13) by Oldrich Tyl, dating from 1929. Glass tiles were used for the first time in its roof. The galleries can be reached by a double spiral of staircases. Unfortunately, the arcade has lost some of its welcoming appearance due to lack of maintenance.

Towards the end of the 20s the modern architectural style of steel, glass and smooth surfaces achieved widespread recognition in Prague as it did in other places, even if not all these "modern" buildings are still admired today. Notable precursors of this pre-Munich period are the **Hotel Julis** (14) dating from 1933 and the **Lindt House** (15) dating from 1927.

Post-war Architecture

The architecture of the post-war years created few buildings that are still admired today. One exception is the **MAJ Department Store** (16), designed in 1968 by the architects' collective Stavoprojekt Liberec. With the generous dimensions of its escalator and its acknowledgement of modern building materials, the building has attracted much admiration and respect both at home and internationally.

Left, apartment house in the Neclanova. Right, escalator in the MAJ department store.

AROUND PRAGUE

There are a number of interesting places around Prague that are well worth a visit. Many of them are close enough for a day trip. One of the most famous places is **Castle Karlstein**, about 21 miles (35 km) from Prague in the direction of Plzen. Castle Karlstein (Karlstejn) was not intended as a residence, but served mainly to protect the Bohemian crown and regalia. The foundation stone was laid on June 10, 1348 and the complex was finished within ten years. Of the various halls and rooms that are open to the public nowadays, the **Chapel of the Holy Cross** is perhaps the most remarkable. It is the most opulent room in the building and the walls are richly decorated with 2,450 precious and semi-precious stones.

Another castle, **Castle Krivoklát**, lies in the direction of the E12 highway about 21 miles (35 km) from Prague. Originally this was a little wooden hunting lodge. Charles IV had this extended into a castle especially for his wife Blanche. Nowadays the castle contains a museum with musical instruments and a gallery of paintings. In the nearby village of **Lány** is the grave of the founder of the Czechoslovak republic, Tomas Masaryk. The countryside around Karlstein and around Krivoklát is excellent for walks.

The palace of **Konopiste**, near the town of Benesov, is famous for its extensive collection of weapons and hunting trophies. **Mělník**, some 19 miles (32 km) away from Prague, lies at the confluence of the Vltava and the Elbe. The area is particularly famous for its vineyards.

The park of Pruhonice, on the southeastern edge of Prague, is one of

Preceding pages: transporting beer. Below, Mělník.

the largest and most beautiful parks in the whole of Europe. It covers an area of 494 acres (200 ha.) and its varied landscape contains 7,000 rhododendrons, azaleas and other shrubs. It is particularly well worth visiting in May and June, when the shrubs blossom.

A little out of the way, in Prague 7 (opposite the Zoological Gardens), lies the palace of Troja, a summer palace built for Count Sternberg. It became famous for its opulent interior decoration and its beautiful staircase with its scenes of the battle between gods and Titans.

Also on the outskirts of Prague, namely in Prague 6, is the **Star Palace** (*Hvezda*), whose name derives from its star-shaped ground plan. The Star Palace which originally lay in the royal hunting preserve, was built in 1555/6. In the 18th century it served as a gunpowder store. However, it was restored in 1949-51 and is now a

museum for the works of Mikolas Ales and Alois Jirásek.

On the road to Kladno, some 16 miles (25 km) from Prague, is the village of **Lidice**, which was burned to the ground after the assassination of the "Reich Protector" Heydrich. Today there is a memorial here and a museum.

About 44 miles (70 km) from Prague in an easterly direction lies the very remarkable town of **Kutná Hora**, which once rivaled Prague in its development of civic splendor. Some of the buildings dating from this time have been preserved, among them the Foreign Court, *Vlassky dvur*, which contains the royal mint. Also worthy of attention are the Gothic St. Barbara's Church and the Stone House, *Kammeny dum*, which is considered a masterpiece of medieval stonemason's work. If you're traveling to Prague via **Cheb** (Eger) you should of course make a stop in **Karlovy Vary** (Karlsbad).

Idyllic countryside just outside Prague.

TRAVEL TIPS

GETTING THERE

BY AIR

Ruzyně Airport is 12 miles (20 km) outside Prague. Here, in contrast to the main Prague rail station, Hlavani nadrazi, one can get porters and trolleys. A taxi to the city center will cost 50 Kčs.

CSA Ceskoslovenske Aerolinie:

Prague 1, Revoluční 1 (Kotva) Tel. 2146 (sale of air tickets, reservations)
Prague 1, Revoluční 25 (Vltava) Tel. 21 46, 231 73 95
(Information about flight schedules and bus timetables to the airport)
Both offices are near the underground station Namesti Republiky.

Ruzyně Airport, Prague 6, Tel. 334
Central Information Service, Tel. 367760, 367814

AIRLINES

Aeroflot, Prague 1, Na přikopě 15, Tel. 26 08 62, Airport: 36 78 15
Air Algerie, Prague 1 inta 23, Tel. 26 54 83, 22 57 70
Air France, Prague 1, Václavské náměsti 10, Tel. 26 01 55, Airport: 36 78 19
Air India, Prague 1, Václavské nám. 15, Tel. 22 38 54
Alitalia, Prague 1, Revoluční 5, Tel. 231 05 35
AUA, Prague 1, Pařížská 1, Tel. 231 27 95, 231 64 69
British Airways, Prague 1, Štěpánská 63, Tel. 236 03 53
Interflug (GDR) Prague 1, Široká 12, Tel. 231 09 55

KLM, Prague 1, Václavské náměsti 39, Tel. 26 43 62, 26 43 69, Airport: 34 14 48
Lufthansa, Prague 1, Pařížská 28, Tel. 231 74 40, 231 75 51, Airport: 36 78 27
Pan American Airways, Prague 1, Pařížská 11, Tel. 26 67 47-9
SAS, Prague 1, Štěpánská 61, Tel. 22 81 41, Airport: 36 78 17
Swissair, Prague 1, Pařížská 11, Tel. 231 47 07, Airport: 36 78 09

BY RAIL

Train travelers arrive at Hlavni nadrazi, the main Prague station. National and international tickets can be bought in Prague from Cedok, Prag 1 Na přikopě, in Western currency, or direct at the railway stations for Czech crowns.

It is wise not to expect porters or trolleys when arriving at the main station. A taxi or the underground can be taken to the hotel. The entrance to the underground is in the ticket hall of the station. The hotel 'Esplanade', Praha 1, Washingtonova 19 is directly opposite the main station and within walking distance.

BY ROAD

A current driving licence and green card for insurance are required. On the visa application one needs to state that a vehicle will be taken.

Lead free petrol (pohonné látky bezolovnatych přísad) is available in the following towns: Brno, Budějovice, Karlovy Vary, Piešťany, Ostrava, Plzen, Teplice, Bratislava and in Prague on both sides of the motorway service station in the direction of Brno. **Diesel** can only be obtained with coupons from the exporters TUZEX. These coupons can be obtained at the border crossings from the Státní banka československa and from some branches of this bank inside the country. The coupons can also be obtained abroad through branches of the company Tuzex.

There is no toll for **motorway** driving in Czechoslovakia. In towns, the speed limit is 37 miles (60 km), outside towns it is 56 miles (90 km), on motorways 68 miles (110 km) per hour. There are frequent police radar checks. Fines for exceeding the speed limit go up to 500 Kčs.

There is an **absolute ban on alcohol** for drivers in Czechoslovakia. Offenders risk losing their driving license and criminal conviction. Children under 12 are not allowed to sit in the front seat. Outside towns **seatbelts are obligatory.**

Petrol prices (as of May 1988)
96 oct. Super 9 Kčs per liter
92 oct. Special 8 Kčs per liter
Diesel (nafta) 5.30 Kčs per liter

According to Czech motorists, diesel is of better quality, and not so contaminated, at the motorway service stations.

Petrol coupons can be obtained from Cedok and at the border crossings. The current price for 20 liters of super is about US$15.00, for 20 liters special, US$14.00.

It is advisable to have sufficient petrol when driving at night. Usually, all petrol stations in an area will be closed in the evening. There is not a very wide network for service and repairs in Czechoslovakia; and spares for Western-made cars are very scarce and expensive.

Passport inspections at the border posts tend to be lengthy and thorough, so that there can be long queues during the main travel season.`

TRAVEL ESSENTIALS

VISAS & PASSPORTS

Every visitor to Prague must have a passport valid for at least five months, and a Czech visa. The visa is valid for three months from the date of issue. Its validity cannot be extended. An entry permit is issued for a maximum of 30 days. Extensions can be made by the relevant local passport offices on payment of the required sum in foreign currency.

The Czech government is introducing a number of measures to make traveling easier. A transit visa has been extended from 24 to 48 hours. Visas may also be issued on the spot at the border posts for 'humanitarian reasons'.

Businesspeople, professionals, lorry drivers etc. can get a visa for 90 to 180 days with multiple entry permits.

Every visitor to Czechoslovakia must register with the police within 48 hours. This is not necessary, however, when staying at a hotel. The address of the registration office in Prague is:

Prague 3, Olšanska 2, Veřejna bezpečnost, Monday to Friday.

Visa applications should be addressed to the local Czechoslovakian embassy. In London, the address is: Czechoslovakian Embassy (visa section), 28 Kensington Palace Gardens, London W8 4QY.

The fee for a visa is at present US$32.00 for a single, US$64.00 for a double visa. It normally takes one to two weeks to get a visa. Journalists, clergy, members of the army, police or security forces have to wait five weeks for a visa. This also involves a telex inquiry to Prague which is charged to the applicant.

MONEY MATTERS

The currency in Czechoslovakia is the Czech crown, Kčs (= Koruna) which divides into heller, hal.(=haleř).Coins in use are to the value of 1, 2 and 5 crowns as well as 5, 10, 20 and 50 heller. Banknotes are to the value of 10, 20, 50, 100, 500, and 1000 crowns.

There is no ceiling to the import and export of foreign currency. It is not permitted to import or export Czech crowns.

OBLIGATORY EXCHANGE

Adults must change US$16.50 per day per person; children aged 6 to 15, US$8.25. It is not necessary if a hotel has been booked through a travel agent or when traveling in an organised group.

A credit card holder who spends at least the obligatory exchanged amount is exempt from obligatory exchange. The possession of a credit card should be noted on the visa application and notified on entry,

BLACK MARKET

It is strictly illegal to change money on the black market in Czechoslovakia. A heavy penalty will be imposed on offenders.

BANKS

Obchodni banka, Prague 1, Na přikopě 14, Mon-Fri 7.30am -12noon

Statni banka československa, Prague 1, Václavské nám.42, Mon-Fri 8.30am-1pm

Statni banka československa, Prague 1, Na přikopě 28, Mon-Fri 8am-5pm

Zivnostenska banka, Prague 1, Na přikopě 20, Mon-Fri 8am-5pm

EXCHANGE

Praha Ruzyně Airport
Cedok travel agency, Prague 1, Na přikopě 18, Mon-Fri 8am-4.15pm, Sat 8.15am-12noon

Pragotour, Prague 1, U Obecniho domu 2, Mon-Fri 8am-9.30pm, Sat 9.30am-6pm

You can also change money in all hotels class A* de luxe, A* and B*.

CUSTOMS

The Czechoslovak customs rules are very strict. An information leaflet about customs regulations is enclosed with the visa.

A list of all personal items, with precise descriptions and code numbers, has to be signed by a customs official at the point of entry. This list should then be presented when leaving, and all personal items must also be taken out again.

Goods for the personal use of visitors are duty free, including: 2 liters of wine, 1 liter of spirits, 250 cigarettes or an equivalent quantity of cigars or other tobacco products, 1000 shotgun pellets or 50 bullets for rifles. A gun license issued by the Czech authorities is required for sporting rifles.

Gifts taken for Czech citizens must not be above the value of 1000 Kčs.

The following goods can be exported customs free without an export licence: all goods that were imported for personal use; travel provisions; 2 liters of wine, 1 liter of spirits, 250 cigarettes or an equivalent quantity of tobacco products, items bought in Czechoslovakia up to a value of 1000 Kčs, goods bought with foreign currency in the TUZEX shops and their branches; a receipt must be shown.

It is illegal to export the following goods if bought with Kčs:

All food items, drinks, cigarettes and other tobacco products, cotton cloth, velour, knitwear, lingerie, baby and children's clothes, furs and leather goods, towels and blankets, shoes and leather gloves, leather trimmings, cutlery and crockery, gold and silver products, spare parts for motor vehicles, car tyres, feathers and feather products, and antiques.

An export permit can be acquired for the following items purchased with Kčs: deckchairs made of rubberised textiles, tents, sporting equipment, enamel and aluminium crockery, plumbing and electric installation materials, building materials, implements and tools for artisans and industrial use, bedlinen, rifles and shotguns.

GETTING ACQUAINTED

GEOGRAPHY & POPULATION

Prague is the capital of the Czechoslovak Socialist Republic (CSSR). It lies on the river Vltava, and stretches across seven hills. It is 581-1310 ft (176-397 m) above sea level, and is situated at 50°8' north and 14°32' east.

According to the census of Jul. 31, 1985, Prague has 1.2 million inhabitants, and occupies an area of 199 square miles (497 square km). The river Vltava , which flows through the town for 23 miles (37 km), is 990 ft (300 m) wide and lies on average 594 ft (180 m) above sea level.

The Czechoslovak Socialist Republic has borders with both the Federal Republic and the Democratic Republic of Germany. Its other neighbors are Poland, the Soviet Union, Hungary and Austria. The population of Czechoslovakia is around 15.3 million, of whom some 64% are Czechs, 30% Slovaks. The remaining 6% are made up of some 62000 Germans in Bohemia and Moravia, 580000 Hungarians in southern Slovakia, 55000 Ukrainians and Russians in eastern Slovakia, and 68000 Poles.

CLIMATE

Prague enjoys a fairly mild climate because of its sheltered situation. Average yearly temperature is 9.3°C.Summer: June 17.3°C, July 19.2°C. Winter: December 1.8°C, February 0.5°C. The yearly average rainfall is 487 mm; it rains least frequently in February, and most often in July.

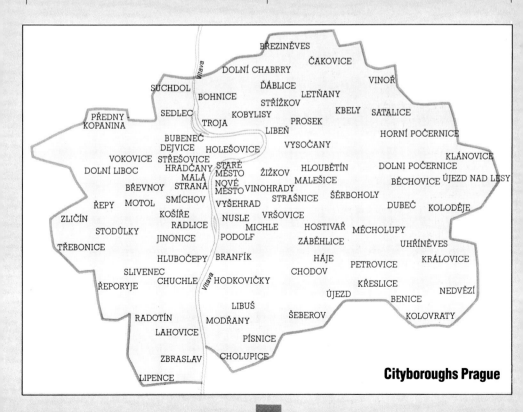

Cityboroughs Prague

COMMUNICATIONS

POSTAL SERVICES

The main post office in Prague is open 24 hours:

Hlavní Posta, Prague 1, Jindřišska 14

Postal Charges		
	Letters	Postcards
Prague & CSSR	Kčsl.(20g)	Kčs 0.50
Other socialist countries	Kčsl. (20g)	Kčs 0.50
Rest of Europe	Kčs 4.(20g)	Kčs 3.00
Austria	Kčs 3.50(20g)	Kčs 2.50
Airmail Overseas	Kčs 6.00(10g)	Kčs 4.40

TELEPHONE

When making a phone call, you can either dial direct or go through the operator. When dialing direct from a public phonebox (if it is in working order) you'll have trouble keeping up feeding five-crown pieces into the machine. The time allowed for 5 crowns is only 4 seconds to Austria, 3 seconds to the Federal Republic of Germany, and 2.5 seconds to Switzerland.

There is a minimum charge of 100 Kčs for a three minute operater-assisted international call. The minimum charge still applies even if the call is shorter.

The cost of a local phone call in the town is 1 Kčs. For local calls through the operator, dial 102 or 108.

EMERGENCIES

MEDICAL SERVICES

There are some surgeries in the University clinic specially for Western visitors. The first consultation is free of charge. Subsequent treatments, operations and stays in hospitals must be paid for in foreign currency. Any medicines prescribed by Czech doctors can be paid for in Kčs.

In an emergency one can turn to the nearest casualty department and to the First Aid posts. There is also a special dental service for foreigners.

MEDICAL SERVICE
FOR FOREIGNERS

Fakultní poliklinika,
Prague 2, Karlovo nám. 32,
Tel. 29 93 81.

—Specialist in internal disorders:
Mon-Fri 8am-4.15pm
—Dentist:
Mon-Fri 8am-3pm.

Emergency Dental Service,
Prague 1, Vladislavova 22,
Daily 7pm-7am
Tel. 26 13 74

Pharmacies (lekárny) open 24 hours:

Prague 1,
Centre, Na přikopě 7,
Tel. 22 00 81, 22 00 82

Prague 2,
Nové Mesto, Ječná 1,
Tel. 26 71 81

Prague 3,
ižkov, Konenova 150,
Tel. 89 42 03

Prague 4,
Nusle, Nám.bratří Synku 6,
Tel. 3 33 10

Prague 5,
Smichov, S.M. Kirova 6,
Tel. 53 70 39

Prague 6,
Břevnov, Pod Marjánkou 12,
Tel. 35 09 67

Prague 7
Letná, Obráncú míru 48,
Tel. 37 54 9

Prague 8,
Leben, Nám.dr.
Holého 15,
Tel. 82 44 86

Prague 9,
Vysočany, Sokolovská 304,
Tel. 83 01 02

Prague 10,
Vršovice, Moskevská 41,
Tel. 72 44 76

Emergency Ambulance
Tel. 333
Emergency Doctor (lékar)
Tel. 155
Traffic Accident
Tel. 24 24 24
Police
Tel. 158

GETTING AROUND

PUBLIC TRANSPORT

TAXIES: It is easy to get a taxi in Prague. An unoccupied taxi will have a lit-up sign. A taxi can be booked for a particular time by telephone through a despatch service. For customers in the city districts 5,6,7 including Malá Strana and Hradčany, the tel. no is 20 39 41; for districts 1,2,3,4,8,9,10, the number is 20 29 51.

Like everywhere else, it is worth making sure that the taxi driver has switched on the meter. A kilometer costs 3.40 Kčs, with a standing charge of 6 Kčs. An hours waiting costs 30 Kčs.

To get to the airport, it is better to take an airport taxi for the CSA, since other drivers will charge for the return journey.

TRAMS

Of all the tram and bus routes within Prague, Route 22 is probably the most interesting for foreign visitors. It goes from Námestí Miru via Charles Square to the highway. It crosses the river over the 1st of May Bridge and passes through the Karmelitská along Malá Strana to the Malá Strana Ring. Then it winds its way up the hill by the castle and carries on through Keplerova, to the departure points for Strahov and Petřín. So you can get a tour of the entire city for 1 Kčs on bus line 22.

UNDERGROUND

Prague is a city which you can explore quite well on foot. If you want to go further, public transport is cheap and the network is well co-ordinated. Central to the Prague public transport system is the Metro. It is

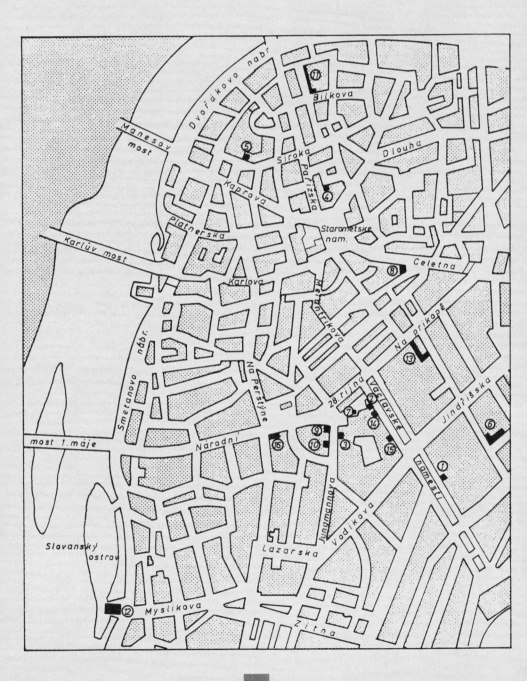

divided into 3 routes. Route A links the east (Leninova), the Malá Strana and the west of Prague (Strašnice). Route B comes from the south, from Smíchov (Smíchovské Nádrazí) via Mustek to the Sokolovská. Here it joins Route C, which runs from Kosmonautu in the south-east to Fučíkova in the north. A single trip costs 1 Kčs.Tickets can be bought either from the machines in the stations or from the little kiosks in front of the station entrances.

PRIVATE TRANSPORT

The type of car available for rental varies from economy, medium and luxury. As such, the cost is between US$29.00 to US$52.00 per day and US$0.27 to US$0.51 per kilometer.

CAR RENTALS

Hertz will rent cars through:

Pragocar, Prague l, Štěpánská 42, Tel. 235 28 25, 235 28 09, Telex: 122 641

Ruzyně Airport, Tel. 36 78 07, Telex: 122 729

Hotel Inter-Continental, Tel. 231 95 95, Telex: 121 353

PARKING ZONES

The city center is kept free of traffic as much as possible. For this reason, the center has been divided into Zone A (Old City), B (New City to the east of Wenceslas Square), and C (west of Wenceslas Square). There are a number of parking zones in the center. Parking fees inside these zones are l Kčs for 3O minutes; in parking zones further outside the center, the fee is l Kčs for an hour.

Parking spaces in the center are:
Platnéřská
Rytířská
Haštalská
Pařížská
Štepánská
Národní
Václavské nam, near the Statni Banka

Námestí Krasnoarmejcu
Na Františku, near Ceskoslovenské Airolinie
Pařížská-Hotel-Intercontinental
Gorkého námestí
Opletalova, near the main station
Politickych veznu
Malá Štepánská
Tešnov, near the Hotel Opera
Petrské námestí, near the tower of Petrská Vež
Petrské námestí
Sázavská
Ibsenova
Škrétova
Tylovo Námestí, near the Hotel Beránek

Only hotel guests with access permits are allowed to drive in Wenceslas Square and the surrounding streets.

Hotel guests can obtain the appropriate access permit from the hotel reception. It must be fixed to the dashboard so as to be visible. The permit allows you to park your car in the spaces reserved by the hotel or in the hotel's own garage. It only applies, however, to the hotel's own parking spaces.

Cars which have been towed away are kept in Prague 10, Hostivar, Cernostolecka [9 miles (15 km) outside Prague].

WHERE TO STAY

HOTELS

Most hotels in Prague are usually booked during the peak season. It would be best to book rooms several weeks in advance through a travel agent, a tour operator or a representative of Cedok abroad to avoid having to search for accommodation.

Besides Interhotels bookings through travel agents, private accommodation is also available. These can be arranged by Cedok and Pragatour.

Hotels in Prague fall into five categories. The prices (Kčs) given below are rough guidelines.

	Single	Double
A de luxe (*****)	1200	1000
A (****)	650	400
B (**)	400	250
Motel	600 with breakfast	
Boatel	550 with breakfast	

The Inter-Continental, Panorama and Forum Hotels are well above this level.

In addition to these hotels, there are so-called 'Boatels' in Prague. They are houseboats in the B* category.

A* DE LUXE (****)

Alcron, Prague l, Štěpanska 40,Tel. 235 92 96. Restaurant, Bar, Exchange Facilities, Garage, Cafe.

Esplanade, Prague l, Washingtonova l9, Tel. 22 25 52. Restaurant, Cafe, Winebar, Bar, Exchange Facilities.

Forum, Prague 4, Kongresova ul.,Tel. 4l 01 11, Fax: 442/42 06 84 Telex: 122 100 ihfpc

Inter-Continental, Prague l, Náměstí Curieovych, Tel. 28 99. Restaurant, Cafe, Winebar, Bar, Exchange Facilities, Garage.

International, Prague 6, náměstí Druž by l,Tel. 32 10 51.Restaurant, Bar, Exchange Facilities, Cafe, Garage.

Jalta, Václavské náměstí 45, Tel. 26 55 4l. Restaurant, Cafe, Winebar, Bar, Exchange Facilities.

A*(****)

Ambassador, Václavské náměstí 5, Tel. 22 13 51-6. Restaurant, Cafe, Bar, Winebarn, Exchange Facilities.

Olympik, Prague 8, Invalidovna, Tel. 82 85 41-9. Restaurant, Cafe, Winebar, Bar, Exchange Facilities.

Panorama, Prague 4, Milevská 7, Tel. 41 61 11. Restaurant, Cafe, Bar, Exchange Facilities.

Parkhotel, Prague 7, Veletržní 20, Tel. 380 71 11. Restaurant, Bar, Winebar, Bar, Exchange Facilities.

U tři pštrosu, Prague l, Druž ického náměstí 12, Tel. 53 61 51-5. Restaurant, Exchange Facilities.

Zlatá Husa, Prague l, Václavské náměstí 7, Tel. 2l4 31 2O.

B*(***)

Belvedere, Prague 7, Obráncu míru, Tel. 37 47 41. Restaurant, Cafe, Winebar, Bar, Exchange Facilities.

Central, Prague l, Rybná 8, Tel. 231 92 84. Restaurant, Winebar.

Europa, Prague l, Václavské náměstí 25, Tel. 236 52 74. Restaurant, Cafe, Exchange Facilities.

Flora, Prague 3, Vinohradská l2l, Tel. 27 42 50. Restaurant, Cafe, Winebar, Exchange Facilities.

Olympik II-Garni, Prague 8, Invalidovna, U Sluncove, Tel. 83 02 74

Päri ž, Prague l, U Obecního domu l, Tel. 231 20 51. Restaurant, Cafe, Winebar, Exchange Facilities.

B(**)

Ametyst, Prague 2, Makarenkova ll Tel. 25 92 56-9. Restaurant, Winebar.

Atlantic, Prague l, Na Porici 9, Tel. 231 85 12

Savoy, Prague l, Keplerova 6, Tel. 53 74 5O. Restaurant, Winebar.

C(*)

Liben, Prague 8, třída Rudé armády 2, Tel. 82 82 27

Národní dum, Prague 3, Bořivojova 53, Tel. 27 53 65. Restaurant.

Tichy, Prague 3, Kalininova 65, Tel. 27 30 79. Restaurant.

BOATELS

Boatels are moored on the banks of the river Vltava and are used mainly by foreign tourists, particularly bus tours.

Admiral B*, Prague 5, Hořejši nábřež i, Tel. 54 86 85. Restaurant, Bar, Exchange Facilities.

Albatros B*, Prague l, Nábřež i L. Svobody, Tel. 42 60 51. Restaurant, Bar, Exchange Facilities.

MOTELS

Club Motel A*, Pruhonice near Prague, Motorway El4, Tel.75 95 13. Restaurant, Bar, Exchange Facilities.

Stop Motel B*, Prague 5, Plzeňská 2l5a, Motorway El5, Tel. 52 11 98. Restaurant, Bar, Exchange Facilities.

ROOMS TO RENT

Rooms can be rented in Prague through Cedok and Pragotour. Both agencies also arrange private accommodation.

Cedok, Prague l, Nové Mesto, Panská ul. 5, Tel. 236 22 92, Telex: 12235 Cedok, Mon-Fri 9am-l0pm. (April to mid-Nov), Sat 8.30am-8pm, Sun 8.30am-5pm, Mon-Fri 9am-8pm (mid-Nov-March), Sat/Sun 8.30am-2pm.

Pragotour, Prague l, Staré Město, U Obecního Domu 2, Tel. 231 72 81, Mon-Fri 8am-9.3pm, Sat 9.30am-6pm.

CAMPGROUNDS

There are a number of category A and B camp sites in the suburbs and outside Prague. The travel agency will make bookings and reservations.

Autoturist, Prague l Nové Mesto, Opletalova 29, Tel. 22 35 44-9, Mon-Fri 9am-l2pm, lpm-4pm.

CATEGORY A

Caravan (May-Oct), Prague 9, Kbely Mladoboleslavská 27, Tel. 89 25 32. Restaurant, Winebar.

Kotva (May-Sept), Prague 4 U lédáren 55, Tel. 46 17 12. Restaurant.

CATEGORY B

Dolní Chabry (Jun-Sept), Prague 8, Dolní Chabry, Usteká ul.

FOOD DIGEST

WHERE TO EAT

This chapter has a high priority for Prague. Some restaurants, cafes and wine bars have already been discussed in earlier chapters.The following gives other, equally good or recommended venues.

For the visitor, the price category - *skupina* - of an eating place is important. There are four categories.

Prices range from around 12 Kčs for roast pork in skupina IV to about 80 Kčs in a first class restaurant. Prices are officially regulated. The menu usually lists the weight in grams for meat or fish dishes. Preparation and ingredients have to conform to a country-wide standard, and this is also the case for special recipes.

Most restaurants in the first two categories are run and supplied by the state-owned company restaurace a jídelny. There is therefore not necessarily a great variety of dishes despite the many venues in Prague.

Restaurants or inns which only serve meals are fairly rare in Prague. They are mostly combined with a wine bar or a pub serving beer. Eating is a sort of national sport, after the maxim 'You can't take it with you - unless it's inside you'.The list below includes inns from skupina III to first class . The restaurants in the big hotels are not listed.

FOREIGN SPECIALITIES

Budapest (Hungarian food), Prague 1, Vodičkova 36, Tel. 24 61 51

Cínská restaurace (Chinese), Prague 1, Vodičkova 19, Tel. 26 26 97

Indická restaurace (Indian), Prague 1, Štepanska 60, Tel. 236 99 22

Jadran (Yugoslav), Prague 1, Mostecká 21, Tel. 26 58 21

Moskva, (Russian), Prague 1, Na přikope 29, Tel. 26 58 21

Sofia (Bulgarian), Prague 1, Václavské nam. 33, Tel. 22 60 98

Viola Trattoria (Italian), Prague 1, Národní tr. 7, Tel. 26 67 32

HOME COOKING

Jewish Restaurant, Prague 1, Maislova 18

Halali-Grill, Prague 1, Václavské nám. 5, (venison), Tel. 22 13 51

Myslivna, Prague 3, Jagellonská 21, (venison) Tel. 27 62 09

Pelikan, Prague 1, Na přikopě 7, Tel. 22 07 81

Praha expo 58, Prague 7, Letenské Sady, Tel. 37 45 46

Savarin, Prague 1, Na přikopě 10 (in the arcade), Tel. 22 20 66

U Kalicha, Prague 2, Na bojišti 12, (Schwejk dishes), Tel. 29 60 17

U Lorety, Prague 1, Loretánské nám 8, Tel. 53 13 95

Vysocine, Prague 1, Národní tr. 26, Tel. 22 57 73

Valdstejnská Hospoda, Prague 1, Valdštejnské nám. 7, Tel. 53 61 95

Reading, smoking, discussing, and letting the day go by. That's how one could describe life in the coffee houses. The former artistic atmosphere of the time of Kafka and Kisch is not very noticeable, but the coffee houses still offer a lot more than just coffee and cake. They are very different from our coffee bars. A coffee house is still part of the essence of Prague life, even though the coffee, often prepared in ten or more different varieties, is not always that strong.

COFFEE HOUSES

Arco, Prague 1, Hybernská 16

Columbia, Prague 1, Staroměstské nám 15

Evropa, Prague 1, Václavské nám 29

Kajetánka, Prague 1, Hradčany, Kajetánska zahrado

Malostranská kavárna, Prague 1, Malostranské nám 28

Mysák, Prague 1, Vodičkova 31

Obecní dum, Prague 1,
Námesti Republiky
  Praha, Prague 1, Václavské nam 10
  Savarin, Prague 1, Na přikopě 10
  Slávia, Prague 1, Národni 1
  U Zlatého hada, Prague 1, Karlova

The menu is usually in two languages in the first class and hotel restaurants. Nearly all waiters in Prague speak English or German. The following is a short list of the most common food dishes:

bažant
pheasant
biftek
steak
bramborák
potato pancakes
brambory
potatoes
buchty
Bohemian desert
chléb
bread
bily chléb
white bread
černy chléb
dark bread
cukr
sugar
drubež
poultry
fazole
beans
guláš
goulash
houby
mushrooms
hovezi
beef
hovezi pečené
roast beef
hovezi varené
meat soup
hrušky
pear
husa
goose
houska
roll
jablka
apple
jeleni
venison

kachna
duck
kanči
boar
kapr pečeny
fried carp
kapr smaženy
carp in breadcrumbs
kapr vařeny
boiled carp
kapusta
savoy cabbage
kaše bramborova
mashed potatoes
knedlíky
dumplings
knedlíky brambo-rové
potato dumplings
knedlíky chlupaté
raw potato dumplings
knedlíky houskové
 bread roll dumplings
knedlíky ovocné
fruit dumplings
králik
rabbit
krocan
turkey
kuře
chicken
kuře smažené
 roast chicken
kyselé zelé
sauerkraut
ledvinky
kidneys
máslo
butter
merunky
apricots
mrkev
carrots
ořechy
nuts
ovoce
fruit
palačinky
thin pancakes
párky
sausages
pečene
roast
pecino
biscuits
polěvka

soup
polěvka dristková
tripe soup
polěvka zelná
cabbage soup
pstruh
trout
rajčatat
tomato
roštenka
oven roast
ryba
fish
ryže
rice
salat
salad
sardinky
sardines
sekaná
mince meat
sladky
sweet
slany
salty
srnčí
roast venison
štika
pike
šunka
ham
telecí
veal
topinky
toast
třešne
cherries
uzeniny
sausage
vejce na mekko
soft boiled egg
vejce na tvrdo
hard boiled egg
vepřová
roast pork
zajíc
hare
zelenina
vegetables
zmrzlina
ice cream
zveřina
venison

DRINKING NOTES

The most important drink in Prague is beer. For the people of Prague, it is the quality of the beer rather than the quality of the inn or its cleanliness that counts.

Many people in Prague only drink their brand of beer and will only go where that is served. Czech beer is probably the best in the world and very drinkable indeed. But to get a hangover you would need to drink quite large amounts.

BEER CELLARS

Branicky slípek, Prague 1, Vodičkova 26, (14 degrees beer from Branik)

Cerny Pivovar, Prague 2, Karlovo nam. 15, (12 degrees Pilsen)

Na Vlachove, Prague 8, Rudé armády 217, (12 degrees Budvar)

Plzenski dvur, Prague 7, Obráncu míru 59, (12 degrees Pilsen)

Smichovsky sklipek, Prague 1, Národní 31, (12 degrees beer from Smichov)

Rakovnická pivnice, Prague 5, S.M. Kirova 1, (12 degrees Bakalar beer from Rakovnik)

U Bonapartu, Prague 1, Nerudova 29, (12 degrees beer from Sichov)

U Cerného vola, Prague 1, Loretánska nám 1, (12 degrees beer from Velké Popovice)

U dvou koč, Prague 1, Uhelny trh 10, (12 degrees Pilsen)

U dvou srdcí, Prague 1, U lužického semináře 38, (12 degrees Pilsen)

U Fleku, Prague 1, Křemencova 11, (12 degrees dark Flek-beer)

U Glaubicu, Prague 1, Malostranské Nám 5, (12 degrees beer from Smichov)

U Medvíku, Prague 1, Na Perštyne 7, (12 degrees Budvar-beer)

U Pinkasu, Prague 1, Jungmannovo nam. 15, (12 degrees Pilsen)

U Schnellu, Prague 1, Tomášská 2, (12 degrees Pilsen)

U Sojku, Prague 7, Obráncu miru 40, (12 degrees Pilsen)

U Supa, Prague 1, Celetná 22, (14 degrees dark special from Branik)

U Svatého Tomáše, Prague 1, Letenska 12 (12 degrees dark beer from Branik)

U Vejvodu, Prague l, Jilská 4,
(12 degrees beer from Velke Popovice)
U zlatého tygra, Prague l, Husova 17,
(12 degrees Pilsen)

Wine bars (vinárna) are at least as popular as the pubs. They mostly serve local wines. The best wine is from Zernoseky in the Elbe valley, which also grows the Melnik wines. Other good wines are from southern Moravia, for instance from Mikulov, Hodonín, Znojmo or Valtice. Some wine bars get deliveries from cooperatives. There used to be vineyards in Prague itself, and the name of the Vinohrady district in the city is a reminder. You can drink foreign wines from Yugoslavia or Bulgaria in the various restaurants serving specialities from those countries. The Trattoria Viola serves *Chianti Ruffino.* The wine bars also serve small meals. Some wine bars, particularly in the Malá Strana, have recently been changed into expensive restaurants and have lost a lot of their former charm.

WINE BARS

Beograd, Prague 2, Vodičkova 5,
(Yugoslav wines,Restaurant)
Fregata, Prague 2, Ladova 3
Klášterni vinárna, Prague l, Národní 8
Lobkovická vinárna, Prague l, Vlašská 17, (Melnik wines, *Ludmilla* restaurant)
Makarská, Prague l, Malostranské nám. 2, (Yugoslav wines)
Melnícká Vinárna, Prague l, Národní tr. 17, (Melnik wines)
Slovácke Vicha, Prague l, Michalská 6, (Bzenec-wine)
U Golema, Prague l, Maislova 8
U Labutí, Prague l, Hradčanské nám. ll, (southern Moravian wines)
U maliřu, Prague l, Maltézské Nám. ll, (southern Moravian wines from Mikulow)
U markyze, Prague l, Nekánanka 8
U mecenáše, Prague l, Malostranské nám. l0
U patrona, Prague l, Dražického nam. 4
U Rudolfa II, Prague l, Maislova 5
U šuteru, Prague l, Palackého 4, (wines from Bzenec)
U zelené žáby, Prague l, Uradnice, (wine from Velke Zernoseky)
U zlaté hrušky, Prague l, Hradčany, Novy svet 3

U zlaté konvice, Prague l, Melantrichova 20

All the wine bars listed above are in the II and III price category. They are traditional Prague wine bars in the Old and New Towns. Opening times vary, most bars are open to midnight, and some even till 3am. As such, they are often the last port of call for Prague late night revelers.

Below is another short list of the most common drinks in Prague:

becherovka
liqueur
čaj
tea
cerny čaj
black tea
džus
fruit juice
káva
coffee
černá káva
Turkish coffee
káva s mléken
coffee with milk
videnská káva
coffee with cream
limonáda
lemonade
pivo
beer
malé pivo
a small beer
černé pivo
dark beer
svetlé pivo
light ale
točené pivo
draught beer
slivovice
slivovitz
sodovka
soda water
vino
wine
bílě vino
white wine
červené vino
red wine
voda
water

THINGS TO DO

CITY

There are many activities in Prague. A tour of the beautiful sights of the city could include a visit to the Zoo, The Botanic Garden or The Observatory.

Prague Zoo is in Prague 7, Troja, and can be reached by underground line C to Fucikova and then by bus 112. A visit to the zoo, which is beautifully kept, could conveniently be combined with a visit to Troja Castle, since they are next to each other.

Prazká zoologická zahrada, Prague 7, Troja, U Trojského mostu 3. Open from February to December, 7am-5pm, in the summer till 7pm.

The Botanic Garden, with its tropical house, is part of the university. It is in the New Town, not far from the church of St. Nepomuk on the Rock.

Botanicka zahrada, Prague 1; Na s lupi, opening hours 7am-7pm.

The **Prague Observatory** is on Petrín hill next to the cable car station. It is open every day except Mondays.

Prague 1, Petrin 205, Jan, Feb, Oct, Nov, Dec 6pm-8pm; March, Sept 7pm-9pm; Apr, Aug 8pm-10pm; May, Jun, Jul 9pm-11pm.

TOUR GUIDES

Cedok, the official Czech tourist office, offers a number of tours around the town and other events. Tours of the town and 'Prague at Night' with a visit to the Alhambra Revue are offered throughout the year. From 15 May to 15 October, there is also a visit to 'Prague at Night' and between June and October, a 'Prague Party' is part of the program. In addition, there are day trips to the environs of Prague during the main season, from 15 May to 15 October.

G O Karlovy Vary-Lidice
G 1 Castles and chateaux of Central Bohemia
G 2 Beauty spots of Southern Bohemia
G 3 Pearls of Czech Gothic Art
G 5 Slápy-Konopište
G 8 Bohemian wine growing regions
G 9 Czech garnet jewelry and "Czech Paradise"

The buses leave outside Cedok, Prague 1, Bilkov 6, opposite Hotel Inter-Continental.

Tickets are available from the hotel reception and from:
Cedok, Prague 1, Bílkova 6
Cedok, Na přikopě 18
Cedok, Ruzyně Airport, Prague

Private guided tours around Prague can be arranged by:
Pražká informační služ ba (Prague Information Service), Prague 1, Panská 4, Tel. 22 43 11

Guided tours around the Hradčany are arranged by:
Informační stř edisko praž ského hradu, Prague 1, Hradčany Vikarská 37 (on the northern side of the St. Vitus cathedral), Tel. 01/33 68

CULTURE PLUS

MUSEUMS

This book is dedicated to the greatest museum in Prague, the city itself. There are, in addition, a large number of museums whose collections and treasures are partly described in the section on tours of the city. This is a selection of various museums.

Architecturally, the **Národní muzeum** (National Museum) is nowhere near as interesting as its predecessor, the National Theatre. It is a somewhat clumsy and solid building on Wenceslas Square. It contains mainly exhibits from the natural sciences, for instance a considerable collection of minerals. Statues from the Libussa Mythology by Ludwig Schwanthaler are in the foyer. The museum has a huge library with over a million books.

Národní muzeum, Prague 1, Václavské nám, Mon-Fri 9am-4pm, Wed,Thu,Sat,Sun 9am-5pm, closed Tue.

If you are interested in technical instruments, measuring implements, and also the first Czech car, the Koprivnitz 'President' of 1897, the Národní technicke muzeum is a must. It contains exhibits of automotive and locomotive construction, photographic items, and sextants from the 16th century in the section on astronomy which were once used by Kepler.

Národní technicke muzeum, Prague 7, Letna, Kostelni 42, Tue-Sun 9am-5pm, Museum archives Wed-Thu 9am-5pm, Museum Library Mon-Fri 9am-4pm.

All kinds of weapons are exhibited in the two military museums. The **Military His-**tory **Museum** in the Schwarzenberg Palace exhibits many unusual weapons, uniforms and war materials from past centuries. It is one of the largest exhibitions in Europe.

Vojenské historické muzeum (Military History Museum), Schwarzenbersky palác, Prague 1, Hradčanské nám, May-October, Mon-Fri 9am-3.30pm, Sat-Sun 9am-5pm.

The creation of the Czech people's army, battles in both World War I and World War II, the resistance etc. are shown at:

Vojenské muzeum, Prague 3, Žižkov, U Památniku 2, Tue-Sun 9.30am-4.30am.

Ethnic exhibits as well as technical implements from those countries are held in the Museum of Asian, African and American Cultures which goes back to the private museum of Vojta Naprstek in 1862:

Náprstkovo muzeum, Prague 1, Betlémské nám. 1, Tue-Sun 9am-5pm, Tue 9am-6pm.

At the bottom part of Petřin is the former palace of the aristocratic family Kinsky, which today holds the Museum of Folklore. It has pottery, glass, toys and national costumes and furniture from old farmhouses. Next to the museum is the beautiful bell tower from Walachia and a little wooden Orthodox church from the 18th century which was transported here in 1929 from Mukačecvo in the Western Ukraine.

Národopisné muzeum, Prague 5, Petřinské sady 98, Tue-Sun 9am-6pm.

The **Craft Museum** has Bohemian glass from different epochs and lovely old furniture. It probably has the largest glass collection in the world as well as a specialist library with around 100000 books which are open to the public.

Umeleckoprumyslové muzeum, Prague 1, Ulice 17, Listopadu 2, (opposite the former Rudolfinum, now the 'House of Artists'), Tue-Sun 10am-5pm.

Philatelists and others interested in post and telecommunications should definitely go to the Museum of Posts which has a large collection of European postage stamps:

Poštovni muzeum, Prague 5, Holečkova 10, Mon-Fri 8am-2pm, Sat and Sun (strictly by appointment).

The **Museum of the City of Prague** is by the underground station Sokolovska. In addition to many exhibits about the history of the city, it holds a rare collection about the history of guilds. But the main attraction is always the famous model of the city, dating 1826-34. You can compare the detailed model of the town from 150 years ago with the Prague of today.

Muzeum hlavního mesta Prahy, Prague 8, Nové sady J. Švermy, Tue-Sun 10am-5pm.

The Strahov Gospels date from the 9th century and consist of 218 handwritten sheets, though comparisons in handwriting suggest that they originate from Trier in West Germany. This manuscript and a number of other very interesting exhibits can be found in the library at the Monastery von Strahov in the Philosophy and Theology Hall. The neighboring rooms in the Monastery houses the **Museum of Czech Literature** which has a collection of some 50000 documents, including letters written by Jan Hus. There is also an interesting reconstruction of a book printing machine from the 17th century.

Památník návodního písemnictví, Prague 1, Strahovské nadvorá 132, Tue-Sun 9am-5pm.

Prague is not only a city steeped in architecture and literature, but also in music, as shown by its museums. There are four main attractions for music lovers in Prague, and you shouldn't miss out on any of them. They contain precious materials from famous composers and musicians.
The **collection of ancient musical instruments** in Prague is famed as the second

largest in the world. In addition to historical musical instruments—which are again being made with the old techniques—it has a rich, majestic collection of written music from various archives.

Muzeum hudebních nástroju, Prague 1, Lázenska 2, Sat,Sun 10am-12pm, 2pm-5pm, weekdays by arrangement.

The name 'Villa America' was given to a lovely small building which houses the **Anton-Dvořak Museum**. It was named after a restaurant in the 19th century, and was built by Kilian Ignaz Dientzenhofer between 1717 and 1720 for Count Michna. Its exhibition rooms contain manuscripts, documents, photos and letters to well-known personalities such as Johannes Brahma or the Conductor Bülow.

Muzeum Antonína Dvořaka, Prague 2, K Karlovu 20, Tue-Sun 10am-5pm.

The **Smetana Museum** is housed in the former waterworks on the river bank in the Old City, and illustrates the life and work of the great Czech composer with manuscripts, written music and other documents.

Muzeum Bedřicha Smetany, Prague 1, Novotněho lPvka 1, Mon-Sun 10am-5pm, closed Tue.

Music lovers will know that Prague is also associated with Wolfgang Amadeus Mozart, for example his *Prague Symphony* and *Don Giovanni*. Mozart lived in the **Bertramka** which now houses the Mozart Museum. The harpsichord and piano in the villa, which is in a lovely area, are still the same ones Mozart used when he was composing his music. The furniture dates also almost entirely from his time.

Muzeum ceske hudby, expozice W.A. Mozarta, Prague 5, Smichov, Mozartova 15, Mon-Fri 1pm-5pm, Sat and Sun 10am-12pm, 1pm-4pm.

The various sections of the **Jewish State Museum** are in the synagogues in the old Prague Ghetto. It is sadly ironic that it was in fact the Nazis, who exterminated 90% of Prague's Jews, who planned to build a

'Museum of Jews as an extinct race', and who collected cultural, artistic and religious artifacts from all over the country. They thus laid the foundation for the Jewish State Museum in Prague, founded by a decision of the Czech government in 1950. The exhibits, which include religious and cultural artifacts, are housed in the different synagogues. The Pinkas-Synagogue will for some time be closed because of renovation work. The entrance fee for all places including the cemetery is 5 Kčs.

One of the most significant memorials to the workers' movement and the Communist Party of Czechoslovakia is the **Lenin Museum,** where the VI Congress of the Social Democratic Party of Russia was held in 1912 under the chairmanship of V.I. Lenin.

Muzeum V.I. Lenin, Prague, Hybernská 7, Tue-Sat 9am-5pm, Sun 9am-3pm.

Opposite the Tyl-Theater in the Rytirska is the Museum named after the communist Prime Minister Klement Gottwald. It contains comprehensive documentation about the working class movement and the Communist Party of Czechoslovakia.

Muzeum Klement Gottwald, Prague 1, Rytiřská 29, Tue-Sat 9am-5pm, Sun 9am-3pm.

ART GALLERIES

The center piece of museums is the **Narodní galerie** (National Gallery)which no art lover should miss. It has seven collections housed in various buildings.

COLLECTIONS IN THE NATIONAL GALLERY

The collections of **Ancient European Art and French Art in the 19th and 20th century** are in the **Sternberg-Palais**, Prague 1, Hradčanské nam. The rooms of the former palace contain such unequaled treasures as the *Rosary Celebration* by Durer, fragments of an altar-piece by Lucas Cranach, and *The Martyrdom of St. Florian* by Albrecht Altdorfer. Amongst the Italian paintings are *David with Goliath's Head* and *St. Jerome* by Tintoretto, the *Portrait of a Patrician* by Tiepolo or the *View of London* by Canaletto. The masterpieces in the Dutch collection include *Haymaking* by Pieter Brueghel the Elder, *Winter Landscape* by Pieter Brueghel the Younger, The *Martyrdom of St. Thomas* by P.P. Rubens and the Portrait of a *Rabbi* by Rembrandt.

Some of the French masters of the 19th and 20th century to be noted are Delacroix, Renoir, van Gogh *The Green Corn Field,* Rousseau, Cezanne and Paul Gauguin. Collections containing works by Chagall and Picasso are also worth visiting.

The third collection is in St. George's Monastery in Jiřsky klástěr, Prague 1, Hradčany, Tue-Sun 10am-6pm. It houses **Old Czech Art,** including works by the Baroque artists Karel Škréta and Jan Kupečkyu.

The **collection of modern art** is in the Mestka knihova (state library) in Prague 1, Staré Mesto, nám. primátora Vacka 1, Tue-Sun 10am-6pm.

The **collection of graphic art** in the Kinsky Palace contains Czech, Slovak and foreign prints of the last five centuries: Palac kinsky, Prague 1, Staromestská nam. 12 (irregular opening hours).

The **collection of Czech sculptures of the 19th and 20th century** is located outside Prague at Zbraslav Castle, Tue-Sun 10am-6pm (April to November).

The renovated convent Anežky kláster, Prague 1, U milosrdnych 17, houses **Czech Paintings of the 19th century** with works by the brothers Quido and Josef Mánes, Karel Purkyne and Mikolaš Aléš.

EXHIBITIONS OF THE NATIONAL GALLERY

Royal Belvedere Palace, Prague 1, Chotkovy sady, Tue-Sun 10am-6pm.

Riding school in the Valdštejnská Palace, Prague 1, Valdštejnská 2, Tue-Sun 10am-6pm.

The **Prague Gallery** has exhibitions held at the:

Old Town Hall Cloisters, Prague 1, Staromestské nám, open daily 9am-5pm.

Old Town Hall, 2nd floor exhibition room, Tue-Sun 10am-5pm.

The **Collection of Rudolf II** was one of

the most important art collections in 16th century Europe. The collection was seriously reduced as a result of looting after the Battle of the White Mountain, annexation by the Habsburgs and Sweden during the Thirty Years' War and through numerous auctions. The remainder of the collection was discovered in the 1960s during a search of the castle's rooms, amongst them paintings by Rubens and Tintoretto. Today, the collection, which has been partly restored, is housed again in the castle gallery, and includes *The scourging of Christ* by Tintoretto, *Gathering of the Olympian Gods* by Peter Paul Rubens. The place to visit is at Prague 1, Hradčany, second courtyard, Tue-Sun 10am-6pm.

The monthly digest of the Prague Information Service also lists exhibitions in the smaller galleries, and the visitor can find out about current exhibitions.

OTHER EXHIBITIONS

Mánes, Prague 1, Gottwaldovo nábřež i 1, Tue-Sun 10am-6pm.

Galerie Nová Sín, Prague 1, Voršilska 3, Tue-Sun 10am-1pm, 2pm-6pm.

OTHER GALLERIES

Vaclav Špála Gallery, Prague 1, Národní tr.. 30, Tue-Sun 10am-1pm, 2pm-6pm.

Gallery Nová sín, Prague 1, Voršilská 3, Tue-Sun 10am-1pm, 2pm-6pm.

Gallery Fronta, Prague 1, Spálená 53, Tue-Sun, 10am-1pm, 2pm-6pm.

Gallery U Rečickych, Prague 1, Vodičkova 10, Tue-Sun 10am-1pm, 2pm-6pm.

Jaroslav Franger Gallery, Prague 1, Betlemské nám. 5, Tue-Sun 10am-1pm, 2pm-6pm.

Gallery bratri capku (Brothers Capek Gallery), Prague 2, Vinohrady, Jugoslávská 20, Tue-Sun 10am-1pm, 2pm-6pm.

Gallery D., Prague 5, Matoušova 9,Tue-Sun 10am-1pm, 2pm-6pm.

Gallery Vincenc Kramar, Prague 6, Ceskoslovenske armády 24, Tue-Sun 10am-1pm, 2pm-6pm.

Gallery ULUV, Prague 1, Národní tr. 36, Tue-Sun 10am-1pm, 2pm-6pm.

Gallery zlata lilie, Prague 1, Malé Námestí 12, Tue-Sun 10am-1pm, 2pm-6pm.

THEATERS AND CONCERTS

Prague has a rich cultural life. More difficult than choice is getting tickets, mostly done by Cedok, though bookings need often to be made days in advance. Even then it is not always certain that tickets are available. That leaves trying the box office, which may well list all showings as vyprodáno - sold out. Don't be put off by this. Friendliness and a hint that you've specially come a long way, and a little something may well get you a ticket.

Some performances are reserved for employees of a company or a cooperative. It's still worth going there and having a look around. Without doubt someone will ask whether you want a ticket, of course for a little extra money. Sometimes you can get help from the usherette.

Advance tickets for concerts and sports and other social events can be bought directly in the Sluna-ticket offices:

Sluna, Prague 1, Panská 4, Passage Cerná ruze

Sluna, Prague 1, Václavské námestí 28, Alfa Passage

Sluna, Prague 1, Panská 4 (cinema tickets)

Sluna, Prague 1, Václavské námestí, Lucerna Passage (rock concerts)

The box office for the National Theatre and the New Theatre is in the Glass Palace of the New Stage, Národní tridá 4 (Mon-Fri 10am-6pm, Sat and Sun 10am-12pm). Tickets for the opera cost between 20 and 100 Kčs.

Národní divadlo (National Theatre), Prague 1, Národní třida 2

Národní divadlo-Nová scéna (New Theatre), Prague 1, Národní třida 4

Smetanovo Divadlo (Smetana Theatre) Prague 1, Vitezněho února 8

Tylovo divadlo, temporarily closed (former Theatre of the Bohemian Estates), Prague 1, Zelezná 11

Divadlo pantomimy-Branik (Mime), Prague 4, Branická ulice 6

Divadlo na Zábradí (Theatre at the railing), Prague 1, Anenské námestí 5

Divadlo Špeijbla a hurvinka (puppet theater), Prague 2, Vinohrady, Rímská 45

Laterna Magica, Prague 1, Národní

třída 40

There are also a number of smaller and less well known theaters. You can find their addresses and programs in the monthly leaflets of the Prague Information Service.

CONCERT HALLS

Dvořák-Hall, House of Artists, Prague l, Námestí Krasnoarmejcu

Smetana Hall, Obecní dum (House of Representatives), Prague l, námestí Republiky 5

Janáček Hall, Club of Composers, Prague l, Besední 3

Palác kultury, Palace of Culture, Prague 4, 5. kvetna 65

Agnes-Areal, Prague l, U milosrdnych l7

CHURCH CONCERTS

Kostel sv. Jakuba, St. Jacob's Church, Prague l, Jakubská.

The Prague International Music Festival 'Prague Spring' takes place each year between l2 May and 4 June. Concerts are then also held in historic halls and churches:

St. Vitus Cathedral, Hradčany
St. Jacob's Church, Prague l, Jakubská
St. Nicolas Church, Prague l,
Malostranské nám
Bertramka, Prague 5, Mozartova l69
Martinic-Palais, Prague l, Hradčanské nám.8
Waldstein-Palais, Prague l,
Valdštejnské nám.

In the cultural centers such as Malostranská Beseda, there is a daily program of musical and artistic events which range from chamber music and jazz to rock concerts.

MOVIES

Few people will come to Prague to go to the cinema. However, there are always interesting films showing. Films such as *Bony A Klid* (about black market exchange touts) are very popular in Prague. You need to queue to get a ticket. In addition to films from the Soviet Union and other socialist states, there are also films from the West, such as those by Ingmar Bergman, Woody Allen etc.

CINEMAS

Alfa, Prague l, Václavské nam 28
Blanik, Prague l, Václavské nam 56
Hvezda, Prague l, Václavské nam 38
Jalta, Prague l, Václavské nam 43
Letka, Prague l, Václavské nam 4l
Lucerna, Prague l, Vodičkova 36
Pariz, Prague l, Václavské nam 22
Sevastopol, Prague 1, Na příkopě 3l
64-U Hradeb, Prague l, Mostecká 2l

Cinema buffs will be interested in visiting the Barandov Studios, which have achieved world fame mainly through their productions of films for children and youth, like *PanTau*. Western producers also work at the Branadov Studios.

NIGHTLIFE

Prague's night life is not exactly wild. While in other European towns insomniacs spend the night living it up, those in Prague presumably go to bed earlier than anywhere else. What is there to do if theaters, operas and wine bars have lost their attractions? Not much, really. However, for those who do want to find out, here are some addresses.Entry fees to get into these bars or pubs are between 30 Kčs and 50 Kčs, the bars come into the I skupina or first class.

One popular place is the **Alhambra** with its nightly show (music, black theater and the usual variety acts). Entry fee is 50 Kčs.
Alhambra Nightshow, Prague 1, Václavské nám. 5, Tel. 22 04 67, 8.30pm-3am.

The **Lucerna Palace** also offers variety shows. The Lucerna Bar is one of the biggest bars with dancing. It also has concerts.
Lucerna Bar, Prague 1, Stepánská 61, Tel. 235 08 88, 8.30pm-3am.

The Park Club is open till 4am on Saturdays and Sundays. It is a favorite venue for businessmen in the Park Hotel.
Park Club, Prague 7, Veletržní 20, Tel. 380 71 11, 8.30pm-3am, Fri-Sat till 4am.

The **Est-Bar** in the Hotel Esplanade has a select atmosphere. The night club, with its changing programs, has the same good reputation as the hotel.
Est-Bar, Prague 1, Washingtonova 19, Tel. 22 25 52, 9pm-3am.

Two bars, the **Jalta Club** and the **Jalta Bar** are located in a luxury hotel, the **Hotel Jalta**. They have orchestras, variety shows and discos.
Jalta Club, Jalta Bar, Prague 1, Václavské Nám 45, Tel. 26 55 41-9, 9pm-3am.

One of the best programs of entertainment in Prague's night life is offered by the **Interconti Club** in the Hotel Inter-Continental.

Interconti Club, Prague 1, Námestí Curieovych, Tel. 28 9, 9pm-4am.

The **Tatran-Bar** also has a good program, including a dance hall with a glass floor.
Prague 1, Václavské nám. 22, 8.30pm-4am, Dancing 5pm-12 midnight.

You can see traditional Czech cabaret in **U Fleku.** Tue-Sat from 7.30pm on. But it is lacking the punch of former Czech cabaret.
U Fleku, Prague 1, Křemencova ul.11

OTHERS

Alfa, Prague 1, Václavské nám. 28, 6pm-1am

Astra, Prague 1, Václavské nám. 4, 10am-midnight

Barbara, Prague 1, Jungmannovo nám. 14, 8.30pm-4am

T-Club, Prague 1, Jungmannovo nám. 14, 8.30pm-4am.

SHOPPING

SHOPPING AREAS

Shopping in Prague can be fun despite the sometimes limited choice of goods on offer. Below are the most important addresses of the Tuzex-shops, department stores and shops in the center.

PRAGUE 1

Rytírska 13 (Gold, watches, jewelry)
U pujcovny 10 (mail order)
Lazarská 1 (women's and men's clothing)
Stepánská 23 (women's and men's clothing)
Palackého 13 (cloth, bed linen)
Ve Smeckách 24 (cosmetics)
Na přikopě 12, Bohemia Moser (glassware, porcelain)
Reznická 12 (Benetton)
Skorepka 4 (car accessories)
Spálená 43 (batteries for digital watches)
Jungmannovo nám. 17 (musical instruments)
Zelezná (cameras)

PRAGUE 2

Francouzská 26 (car accessories)
The main department stores are:
Prior, Bílá labut, Prague 1, Na Porici 23
Prior, Kotva, Prague 1, Nám. Republiky 8
Prior, Detsky dum, Prague 1, Na přikopě 15
Prior, Máj, Prague 1, Národní tr. 26
Druzba, Prague 1, Václavské nám 21
Dum kozesin, Prague 1, Zelezná 14
Dum mody, Prague 1, Václavské nám 58
Dum obuvi, Prague 1, Václavské nám
Dum potravin, Prague 1, Václavské nám 59

ANTIQUES

Galerie starozitností

Prague 1, Mustek 3
Prague 1, Mikulandská 7
Prague 1, Národní tr. 24
Prague 1, Václavské nám 60
Prague 1, Melantrichova 9

BOOKSHOPS AND SECOND-HAND BOOKS

Prague 1, Dlázdena ul. 5
Prague 1, ul.28.října 13
Prague 1, Stepánská 42
Prague 1, Vodickova 21
Prague 1, Na přikopě 27
Prague 1, Mustek 7
Prague 1, Staromestská nám 16
Prague 1, Celetná

RECORDS

Prague 1, Václavské nám 17
Prague 1, Václavské nám 51
Prague 1, Vodickova 41
Prague 1, Jindrišská 19
Prague 1, Celetná 8

ARTS AND CRAFTS

It is exhibited and sold in the state-controlled shops 'Dílo'.

Dílo, Prague 1, Na příkopě near underground station Mustek
Dílo, Prague 2, Vodičkova 32
Gallery Centrum, Prague 1, ul.28. Října 6
Gallery Pralyz, Prague 1, Národní tr. 37

GLASS AND JEWELRY

Bohemia Glas, Prague 1, Parízská 2
Bohemia Moser, Prague 1, Na přikopě 12
Borske Sklo, Prague 1, Malé nám 6
Krystal, Prague 1, Václavské nám 30
Bijouterie, Prague 1, Na přikopě 12
Prague 1, Václavské nám 53
Prague 1, Národní tr. 25

SPORTS

PARTICIPANT

There are of course opportunities for sports in Prague. Specially for tennis fans who want to have the chance of being taught in the very successful Czech technique, **Cedok** arranges courses from June to September (individual and group tuition).

Hunting can also be booked through **Cedo**k in the appropriate season.

Riding excursions are offered in Konopiste near Prague and can also be booked through **Cedok.**

ICE SKATING

Slavia Praha IPS Stadium, Prague 10, Vrsovice

Stvanice Sport-Areal, Prague 7, under the Hlavkov most (bridge)

SWIMMING

Swimming in the open air swimming pools by the river Vltava is not recommended because of the state of the water.

INDOOR SWIMMING POOLS

Dum Kultury Klarov, Prague l, Nábřeží kapit.Jarose 3

Julius Fucík Park, Prague 7

Tickets for sporting events can be bought both through **Cedok** and in the central advance booking office, Prague l, Spálená 23.

The state owned horse racing track is in Chuchle, outside Prague.

FURTHER READING

Kafka, Franz:*America/The Trial/The Castle,* translated by W. and E. Muir, Penguin Modern Classics.

Kafka, Franz:*Description of a Struggle and other stories,* translated by W. and E. Muir, Penguin Modern Classics.

Kafka, Franz:*Diaries,* translated by W. and E. Muir, Penguin Modern Classics.

Hasek, Jaroslav:*The Good Soldier Schweik* translated by Sir C. Parrot, Heinemann.

Vaculík, Ludvík:*A Cup of Coffee with my Interrogator,* translated by G. Theiner, Readers International.

Vaculík, Ludvík:*Prague Chronicles,* translated by G. Theiner, Readers International.

Seifert, Jaroslav:*Selected Poetry,* translated by E. Osers, Andre Deuts.

Neruda, Jan:*Tales of the Little Quarter,* translated by E. Pargetwr, Greenwood Press London.

USEFUL ADDRESSES

TOURIST INFORMATION

The Prague Information Service has an office in the city, at Prague 1, Na přikopě20, which has available all information about Prague.It also stocks maps of the town, and booklets called *A Month in Prague* containing useful information and addresses. You can also find the monthly English and German language listings magazine free of charge. This lists plays and concerts, exhibitions and a selection of other cultural events. There is a branch of the Prague Information Service at the underground station Hradcanská (Line A).

The Czech language magazine *prehled kulturních poradu v Praze* gives detailed information about all events, including those in theaters, museums etc. in each of Prague's districts.

Travel information and bookings are in most cases done by **Cedok.**

Section for rail and air travel, Prague 1, Na přikopě18, Tel. 12 11 09, 12 18 09, 12 22 33.

Section for tourist attractions, cultural events, and tickets, Prague 1, Bílkova 6, (close to the Hotel Inter-Continental), Tel. 231 87 69, 231 89 49.

EMBASSIES

Austria, Prague 5, ul.Viktora Huga 10, Tel. 54 65 577, 54 65 50

Canada, Prague 6, Mickiewiczova 6, Tel. 32 69 41

Federal Republic of Germany, Prague 1, Vlašská 19, Tel. 53 23 51

France, Prague 1, Velkopřevorské nám. 2, Tel. 53 30 42

Great Britain, Prague 1, Thunovská 14, Tel. 53 33 47

Italy, Prague 1, Nerudova 20, Tel. 53 26 46

Japan, Prague 1, Maltézské nám. Tel. 53 57 51

Netherlands, Prague 1, Maltézké nám. 1, Tel. 53 13 78

Switzerland, Prague 6, Pevostni 7, Tel. 32 83 19

United States of America, Prague 1, Tržište 15, Tel. 53 66 41

ART/PHOTO CREDITS

INDEX

P

R

S

Y - Z

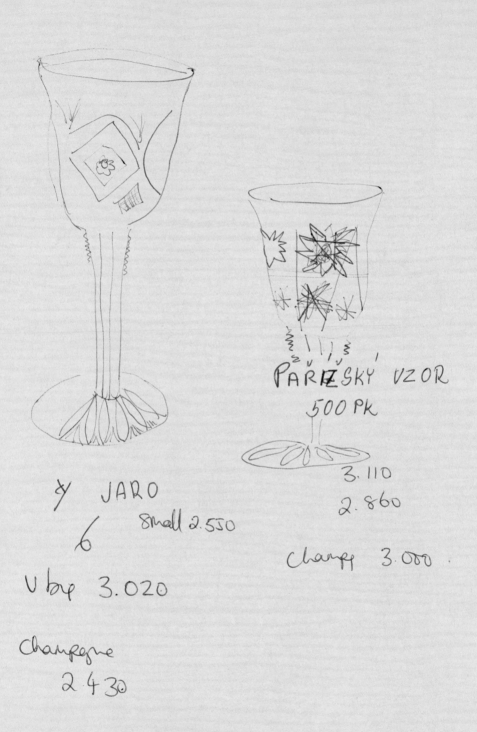

PAŘEŠSKÝ VZOR
500 Pk

JARO

small 2.550

V bye 3.020

Champagne
2.430

3.110
2.860

Champ 3.000